the conduct
of
science

DATE DUE

DEMCO 38-296

the conduct

of

science

MICHAEL W. FRIEDLANDER
Washington University

Prentice-Hall, Inc., *Englewood Cliffs, New Jersey*

10   9   8   7   6   5   4   3   2   1

ISBN: P 0–13–167262–2
*   C 0–13–167270–3*

Library of Congress Catalog Card Number: 72–392

Printed in the United States of America

PRENTICE-HALL INTERNATIONAL, INC., London
PRENTICE-HALL OF AUSTRALIA, PTY. LTD., Sydney
PRENTICE-HALL OF CANADA, LTD., Toronto
PRENTICE-HALL OF INDIA PRIVATE LTD., New Delhi
PRENTICE-HALL OF JAPAN, INC., Tokyo

table of contents

preface

For many years, science has been included among the areas that make up a liberal education at the undergraduate level. It seems to have been agreed that some exposure to science was desirable, as were introductory surveys in literature, philosophy, and a range of other subjects. For the prospective scientists and engineers, the introductory science courses have usually been demanding and rigorous. In contrast, there have grown up, in many science departments, elementary courses specifically aimed at the liberal arts major. These courses have tended to be highly qualitative, often using a bare minimum of mathematics (and then apologetically) and seeing as their objective the explaining of the observed phenomena of the world around in simple terms. In physics, it has been easiest to offer a diluted version of the standard freshman course, but still to require a drill of homework problems as a show of quantitativeness. The overwhelming impression that one gains is that these courses dare not be too demanding for fear of failing too many students, and that even this sort of exposure to science will still serve some useful broadening purpose. Enrollment in many of these courses is maintained because of the science requirement for graduation. Occasionally, one of these survey courses rises far above the general level, through the personality and efforts of a particular instructor, but, unfortunately, all too often they have remained as requirements, something to be "got out of the way," perhaps even mercifully during a summer session.

Within the past few years, there has emerged a recognition that something needed to be done to make the introductory science courses more interesting, and at the same time a better reflection of contemporary scientific thought. Revised curricula have been developed in mathematics, physics, chemistry, earth science, and biology, both for the high school and college levels, and further experimentation will surely continue. For the most part, these courses and their texts develop the rudiments of scientific methodology and then explore a variety of topics that are now considered central to the respective subjects. This is good, but limited. Although the nonscientist may be interested by this approach, he is unlikely to have to use this knowledge later. He is, though, almost certain to be asked to give support to, or to oppose, scientific and technical projects whose technical content he probably does not understand but whose costs, through taxation, he definitely appreciates. He will also probably be increasingly aware of environmental problems, which often turn into political issues.

The obvious answer to this is the introduction of "relevant" material: how a laser works; the physics of the sonic boom that accompanies airplanes travelling at supersonic speeds; or the chemical and biological effects of defoliants and pesticides. Such topics may have a timeliness, and can be used with great effect to enliven a class and illustrate some special points. Underlying any discussion of such topics, however, there should be some understanding of the more permanent workings of science and the scientific community, so that the nonscientists can emerge with a better understanding of why some scientists' opinions can be accepted and others' not, or how unorthodox scientific theories can be challenged and tested, or of the nature of the sciences' internal system of checks and balances. These internal mechanisms of science will change only slowly; an understanding gained now will still be of use many years hence, even though the topics of concern have shifted.

This book is designed to be useful as an adjunct to any introductory science course. All of the topics have been included in a two-semester course that I have given for nonscientists, although some topics have been considerably expanded here. Some formed the basis for class assignments and term papers, followed by extensive in-class discussions. Others arose as I tried to develop particular areas. With two semesters available, it was possible to develop a critical and quantitative approach in introductory physics which could then be carried over to many of these "relevant" topics. On the other hand, my use of some of this material was not as successful in a weekly seminar on "Science and Society," where there was not the central quantitative thread. From this limited experience, my conclusion is that it takes time to cultivate even a minimum understanding of the scientific method, but that this greatly aids the approach to the "relevant" parts of the course. A seminar or course that has only these softer topics with no quantitative foundation lends itself to exploitation by those students who are happy to offer value judgments but seem quite unwilling to undertake the critical reading and thinking necessary for serious discussion.

Many students try to keep away from science, which they wrongly and simplistically blame for many of the present ills they observe in society. This sort of opinion will probably not be altered without at least a basic understanding of the workings of science and its relationship to society, and that is one aim of this book.

As my course developed, I became aware of the difficulty in assembling suitable reading materials. It is this, in particular, which has prompted my present writing. It would, of course, have been possible to have written far more on each of the subjects chosen for the following chapters. What is intended here is rather to indicate some subjects that do interest nonscientists, and to explore them briefly, giving for each some guidance for further reading. It is my firm conviction that much of the benefit for each student comes from doing his own reading and thinking and then writing and discussing with critical comments from the class and instructor. The hardest part is the delineation of topics and knowing where to start. Although I may have strong preferences for certain answers to some of the questions raised, I have tried to indicate where there is no agreed solution or position, and where each person must make his own value judgment. In many places, I shall ask questions, but shall not attempt to provide answers.

The order of the chapters is largely arbitrary, and others may well prefer to proceed differently. Personal preferences may lead to some chapters' being omitted in a course, and other subjects' being inserted. What *is* important to the success of a course is the real interest of the instructor, and the choice of material is clearly a matter of taste. The final result, hopefully, will be the same, in that students will have been interested rather than alienated.

It is a pleasure to acknowledge, at this time, the several sources of assistance towards the writing of this book: my students, in classes and seminars, for their interest; many friends in Washington University and the Committee for Environmental Information, for innumerable discussions and arguments over the years, through which my interest in the social aspects of science has been stimulated and sharpened; and, especially, my wife, not only for her critical and helpful review of the manuscript but also for her shared interest and continued encouragement.

acknowledgements

I wish to acknowledge, with thanks, the permission received from the following publishers to reprint material from their respective books:

The Brookings Institution, for extracts from *Science Policy and the Universities*, edited by Harold Orlans.

Columbia University Press, for extracts from *The Rise and Fall of T. D. Lysenko*, by Zhores A. Medvedev, transl. by I. Michael Lerner, with editorial assistance by Lucy G. Lawrence.

Doubleday and Company, and The Society of Authors on behalf of the Bernard Shaw Estate, for extracts from *Shaw on Music*, edited by Eric Bentley.

Dover Publications Inc., for extracts from *Dialogues Concerning Two New Sciences*, by Galileo, transl. by Henry Crew and Alfonso De Salvio.

Harcourt, Brace, Jovanovich and Jonathan Cape, Ltd., for extracts from *The Notebooks of Leonardo Da Vinci*, edited by Edward MacCurdy.

McGraw-Hill Book Company, for extracts from *Boswell: The Ominous Years*, edited by Charles Ryskamp and Frederick A. Pottle.

The M.I.T. Press, for extracts from *Reflections on Big Science*, by Alvin M. Weinberg.

In addition, permission of editors and authors is gratefully acknowledged for extracts from the following journals:

American Scientist; *The Bulletin of the Atomic Scientists*; *Nature*; *New Scientist*; *Physics Today*; *Scientific American* and W. H. Freeman and Company; *Science*.

For permission to reprint noncopyrighted materials in the appendix, I am grateful to the American Association of Physics Teachers, and Professor Jay Orear, Cornell University.

Explicit references are contained in the appropriate footnotes.

the conduct
of
science

the social direction of scientists

chapter 1

Even a casual glance at daily newspapers or weekly review magazines soon reveals the extent to which science and its applied aspects, technology, comprise an important part of national and international affairs: Concern over pollution of air and water; the use or banning of various weapons; the choice between different means of generating electric energy; the advantages or unexpected side effects of widely used medicines—there seems no end to this list. What these apparently unrelated items have in common is their pervasiveness or their price. Science and technology are no longer minor items in the national budget, nor are they simply the hobby of a few. Perhaps not too long ago, science could be classed with the study of the classics or archeology or philosophy in terms of immediate usefulness. Such times have gone, and in many ways the direction that science takes is determined not by the scientists alone but by the sum of all influences and forces with politics now playing an increasingly important role. This is a major change from earlier times when each scientist could make his own decisions; now whole fields of science prosper or decline as a result of decisions made at the federal budget level. Similarly with technology: Increasingly there is criticism of the way in which the forces of the market place dictate the way in which our natural resources are used, and the many ways in which our way of life depends upon and has been changed by technology. As a result there is a growing pressure to regulate such matters through legislation.

Out of this there has emerged an increased awareness of the potential power of science, but unfortunately the general understanding of the way in which science is conducted has not kept pace. It is certainly unrealistic to expect all of the public to have a deep understanding of all technical and scientific matters, but it is surely not unreasonable to hope that many will have some understanding at least of the ways of the scientific community. In the absence of such understanding, there have in the past been instances of quite unfair criticism of the scientific community (as for instance, in implying that the development of nuclear weapons by Russia and China could only have come via secrets obtained from the United States); in the future there may well be quite impossible expectations of science with subsequent disappointment and resentment.

There is an extensive literature in which the philosophy, the history, and the structure of science are examined. Most of this literature has come from people who are not active scientists—they have mostly been philosophers, historians, political scientists, sociologists, science writers, professional critics, and sometimes devoted amateurs in science. As would be expected, these many different authors have looked at science from different angles, have started with different attitudes and biases, and have seen science in many different lights.

Scientists themselves are no less diverse in their attitudes toward and their understanding of science. To some, science is important for its aesthetic qualities in which the breadth and elegance of its laws are comparable to other creations of human imagination in the arts and literature and thought. To others, science appeals because of the applications that have been responsible for the improved lot of at least some of the earth's population. For still others, science is simply a means of earning a living. Too often though, the scientist is viewed by others as something apart, perhaps semihuman or robotlike or cold, to be caricatured as a man in a white coat, wearing thick-lensed spectacles and speaking with a heavy accent. Stereotypes persist far too long, often turning into objects of prejudice, and the scientific community has at times been viewed by the public at large in highly uncomplimentary ways. Much of this can be traced to a fear—of the forces which science is presumed to control. How can one then entrust the security of this country to such people? Or, more recently, are these not the technological sorcerers to be blamed when the broom backfires in the hands of the apprentice and we find ourselves engulfed in a cloud of polluting dust? Despite these widely held views, scientists are members of the community in many ways: They go to PTA meetings, play golf, mow their lawns, get parking tickets, and in countless ways live normal lives. The differences enter in their professional lives. Science itself is conducted according to its own rules, and associations of scientists have been formed to promote the conduct of science. Now, with the immediate and worldwide impact of scientific and technical advance having become so apparent, attitudes to science are changing. Many scientists are coming to feel obligations to society that go beyond the mere popularization of current knowledge. Society expects more of scientists. An accommodation has not yet been reached, and it requires better mutual understanding.

The "scientific method" has received a great deal of attention. It gets some attention in many high school and college texts and closer examination by philosophers of science. The more detailed treatments have tended to focus on topics such as scientific inference and induction, the validation of scientific theories, and problems of knowledge. Most working scientists, however, show little concern for the "philosophy of science." In their daily science, they are apparently not influenced at all by any explicit knowledge of the methodology so carefully analyzed and described by others.

The internal workings of science, as they are recognized by active scientists, have been very well discussed by Ziman,[1] and it is not my intention to duplicate his coverage. Rather, my main concern is to point to what might be termed the foreign policy of science—its external relations and problems, which may emerge from the application of scientific advances or from other ways in which organized science comes into contact with the larger society.

Although individual invention and imagination are absolutely basic to the progress of science, the game is played according to universal rules. This may sound paradoxical, but the rules have been largely responsible for the growth of science. The universal criteria by which scientists judge each other's work are absorbed during the scientists' training, usually at the graduate level, and become so deeply ingrained as to color the way in which scientists approach political and social problems. Some understanding of this internal structure can greatly illuminate the later discussions, and to that end the subjects of scientific information and authority are treated first. Here as in most of this book, there is no intention of providing an exhaustive treatment. Instead, the discussions should be regarded as introductory. Problems and issues will be identified; sometimes alternative solutions will be discussed, but often the questions remain. Since most of the issues raised are by no means closed, any continued interest can be maintained by following the reports in appropriate publications. For all issues, however, the references and bibliography will provide a starting point for those who might wish to go into greater detail.

Some of the topics that are discussed in later pages are of concern mainly in science, while others have their effects in technology. Although we may all have an understanding of the distinction between science and technology, that distinction is not easily defined in the abstract and with complete generality. Science, we may feel, deals with the expansion of basic knowledge and the uncovering of the rules of nature, whereas technology consists of the application of this knowledge in the cause of human welfare. Such simple differentiation does not long stand up to scrutiny: In order to market some new gadget, it may be found that a particular piece of basic knowledge is needed. Does the search for this basic knowledge now constitute applied research since it has been stimulated by the desire for a specific application? This question will be discussed again later. The main point to be made here is that the social questions that scientists now face are also confronting those who work in technology, and it is usually a matter of style which will dictate in later pages whether the terms scientist and/or technologist are used. Although many problems are common to both groups, there are important differences between them in the freedom they have to respond to these problems.

As the realization has grown within the scientific community of the extent and importance of the problems raised by the application of science and technology, the response has taken two lines. In one, the scientists have been drawn in as advisers. They have advised a variety of government agencies and many have advised the public; that is, they have tried to inform the public of the technical background to the current issues. The other line is just as important: The scientists have recognized the importance and need for having themselves a better understanding of the social institutions that are so affected by the technological advance. Especially in the field of international disarmament and the improvement of contacts that will reduce tensions, the scientists have been active. Even before the first use of atomic weapons, a group of scientists in Chicago had petitioned against the use on centers of population. From this deep concern has grown that influential and fine magazine, *The Bulletin of the Atomic Scientists.* Closely related have been the annual Pugwash Conferences, which now cover a host of subjects and provide invaluable contacts between scientists from East and West. It is sad to record that to a considerable extent, the scholars in the humanities have lagged very badly in their understanding of the revolution that has proceeded around their ears. Some have deplored the rush of progress, but too few have understood. When C. P. Snow raised some very serious questions in his lectures, *The Two Cultures,*[2] the response from a major literary critic was a scorching denunciation of Snow's style and his novels, coupled with a complete inability to answer Snow's main thesis or even a recognition of this omission. There could have been no clearer demonstration of the cultural gap to which Snow was pointing. It is partly in hopes of diminishing the excuse for the persistence of such defiant and ignorant obscurantism that this book has been written.

NOTES AND REFERENCES

1. J. M. Ziman, *Public Knowledge* (Cambridge: Cambridge University Press, 1968).

2. C. P. Snow, *The Two Cultures: And A Second Look* (New York: Mentor Books, The New American Library, 1959, 1963).

spreading scientific information

chapter 2

2.1 Practices, Problems, and Needs

Even in its most radical moments, science remains conservative. The latest theories, which may represent the greatest breaks with prior thinking—for instance, Einstein's theory of relativity—must still draw upon the accumulation of past knowledge. New ideas do not overthrow old facts; they replace older ideas but must still include the known facts within their regions of validity. As science progresses, data and analyses must be recorded and from these will come the hypotheses and theories that will guide the next round of experiments and observations. For science to prosper, scientists must share their information, exposing their results and theories to the critical review of their community. The processes by which the scientific community arrives at its consensus regarding the most widely accepted body of current knowledge are far more subtle than the "scientific method" which is briefly described in introductory texts. Penetrating reviews of these processes have been given by Ziman[1] and Kuhn.[2] Some aspects of this problem will be considered in Chap. 3, but here only one important part of the scientific process will be pursued: The dissemination of scientific knowledge itself.

Why should scientific knowledge be freely available? Would it matter if each scientist were to keep his notebooks locked up, being content to work in his laboratory and talk to no one? We can immediately think of two areas where secrecy does prevail. In the applied sciences involving industrial applications and competition, secrecy is understandable. Despite the protection of patents, some industrial processes are closely guarded, and even greater secrecy attends the development stages that precede the application for a patent. In such a highly competitive arena, progress may be uneven possibly even depending on the accident of the presence or absence of particular skills in the composition of a research team. (It is interesting to speculate whether technological progress might be more rapid were the information freely shared across an entire industry, but perhaps the present combination of secrecy and competition compensates effectively.)

The other area of secrecy involves military technology. Given the existence of international fears and tensions, it seems unreasonable to expect that all military research will be conducted in the open. The very existence of some military research may be hidden from the general public, and the prime example of this was the Manhattan Project for the production of atomic weapons in World War II. The existence of research in many areas may be well-known although the technical details are kept secret, while in addition, the Department of Defense supports a large amount of nonsecret research, both in its own laboratories as well as in universities and industry.

Secret scientific research, whether military or industrial, cannot contribute to the mainstream of the progress of science. Instead it raises serious questions as to where it should be conducted and, in particular, the extent to which universities should be involved. These important questions will be discussed in Chap. 6, but in the remainder of the present chapter, our attention will be confined to the open scientific literature from the professional to the popular.

At an earlier stage in the development of science, word travelled more slowly, and it was also common for one scientist to repeat the experiments he had just come to hear about. For instance, Galileo, in his earliest observations, discovered four moons around Jupiter, using the astronomical telescope he had just invented. Kepler had corresponded with Galileo. Through an intermediary, Kepler was able to borrow a telescope made by Galileo, repeat the observations of Jupiter's moons and then show that the motion of these moons around Jupiter obeyed the law of celestial motion that he had so recently deduced for the orbits of the planets around the sun. Today this procedure is not generally followed. An exception occurs when some results are reported that are so unexpected that they need prompt confirmation, but the volume of information in science is now so large and covers such a range of specialized subjects with unique apparatus that further progress will come only if we usually accept and are guided by the results of others. Of course, this acceptance may be tempered by our regard for the other experimenter and the reputation of the journal in which he has published.

Science needs and thrives on the communication of information. Robert

Hooke, 300 years ago, hid his newly discovered law of elasticity in a crypto-gram, but today's scientists usually err in the other direction, especially in highly fashionable and competitive fields where priority of discovery can bring benefits of prestige and research funds. There is more to be gained, however. With experiments often being costlier and taking much longer to devise, unnecessary duplication can waste precious funds, and experiments in progress can often be improved by the incorporation of changes based on the results of others.

So far we have been following the path of information within the scientific community. In the next section, we shall explore in more detail this spreading of information at the professional level—through journals, conferences, and less formal procedures. Before we start into those topics, however, a brief comment is appropriate regarding some other problems that have received far less attention than they deserve.

It is very useful for scientists to have an awareness of technical advances in other areas of science than their own. Fundamental discoveries and improved methods in one science can often be adapted for use in other areas, and new fields continually open up between the boundaries of conventional and established disciplines. For example, advances in the understanding of the physics of solids at very low temperatures have led to improved detectors of infrared radiation, which are now being applied in astronomy; the applica-tion and adaptation of physical and chemical techniques have transformed the biological sciences. There are at least three routes by which information can cross disciplinary boundaries. First, a scientist may develop a pro-fessional expertise in more than one area. Second and more frequently, scientists may draw their inspiration from popular or semipopular literature, and there are several journals such as the *Scientific American* that can fulfill this function. Third, there can be collaboration or casual contact between scientists. Although all of these routes have been productive, much depends on the interests and personalities of the individual scientists involved.

If we were to go no further, we would have described and defined the boundaries to the interests of only a very small part of the population. Science does not exist in a vacuum. Its results find industrial and military applications; its support can no longer be hidden in the dark corners of the national budget nor left to philanthropy. Large amounts of public money go into projects involving science or its applications: The space program; the building of large particle accelerators for nuclear physics research; a development program for supersonic airplanes; the nuclear energy program. All of this is very much in the public eye, usually with dollar signs associated. (A good measure of the way in which money has come to be associated with science, at least in the mind of some of the press, was the announcement of a major award for contributions to science: "Dr. ——— wins $40,000.") More recently the problems of pollution in the environment have gained very wide publicity. In all of these, the public, through Congress and increasingly through local government too, is being asked to provide via taxation the support needed. It is unreasonable to expect continued and unquestioning support except possibly in times of national emergency. What is desperately needed is a much wider public understanding of the ways of science, the benefits that accrue to society through scientific research and also the problems raised, and hopefully something of the fascination and enjoyment and satisfaction involved in research. It is not easy to popularize all of this, but the scientific community must make far more effort than it has hitherto. With the cooperation of active scientists, we can hope to move from breathless reporting that dwells on the most extreme of the doomsday prophecies or the wonders of science to lively accounts that retain scientific accuracy and integrity. This is not easy, and there is room for more good scientific journalism. Without this to help provide a wider basis of understanding, decisions will of necessity have to be made by smaller groups often with narrower interests. Decisions that are primarily political may have a substantial scientific or technical base, and the distinction must be made between the politics and the science. So, for instance, in the debate over the testing of nuclear weapons in the atmosphere, the scientific debates revolved around the distribution of radioactive fallout and the biological effects that were feared when fallout was ingested by man. With the scientific aspects clearly defined—there was general agreement on some points and no agreement at all on many others for which there was very limited information available—attention could then turn to the politics. Given the scientific knowledge and the remaining uncertainties, were the potential biological hazards sufficiently small that weapons testing should have continued? The other side of the problem was the political assessment of the need for further tests and the anticipated consequences of either testing or not testing. The decision in the end was political and strategic, but no intelligent decision could be made without a clear understanding of the scientific aspects. Here clearly was a situation where the public needed at least a rudimentary understanding of the prospective risks and benefits for the alternatives open. More recently the debate on the antiballistic missile (ABM) has raised the same problems, and so has the concern with environmental pollution and supersonic transport. The public needs scientific information.

Beyond these important issues combining science and public policy, there is room for a wider appreciation of science as a product of human ingenuity on a par with philosophy or literature. The laws of science and their use in the description or explanation of everyday phenomena that would otherwise continue to appear as miracles; the intellectual structure and the elegance of the formulation of the sweeping generalizations that we term laws—all of this deserves a broader recognition.

How does one bring science to the public? Most of the population are not scientists. Many have endured some introductory courses in high school or college to emerge with little understanding or interest. Major efforts since the middle 1950's have led to greatly improved curricula and books and simple laboratory apparatus, and this should be continued. Despite this large effort, science frightens most laymen. Scientists can try to stimulate the public's interest and concern and must then help with information in understandable form. The press, radio, and television have obligations here largely unfulfilled. This topic of popular science will be addressed in the last section of this chapter; its importance should not be underestimated.

2.2 Professional Publications

Until about 300 years ago, scientific communication was sporadic and individual. Over the centuries, discoveries and theories were published in treatises, often in a language or style that effectively precluded a wide readership. Correspondence between scientists was sometimes extensive but could proceed no faster than the mail permitted. A major change took place around the middle of the seventeenth century when groups such as the Royal Society of London were formed. Meetings of scientists and interested amateurs were formalized. Publication of the *Philosophical Transactions of the Royal Society* in 1664 was the start of the professional journal that today caters to specialists and makes no concessions to popular taste. Originally, however, these journals contained all manner of rambling and discursive reports, chance observations, and hearsay in addition to those which we would today classify as scientific. Professional journals are now far more than simply highly technical newspapers or archives; in their present form, they are also the guardians of professional standards representing the controls that science imposes upon itself.

At the very heart of the system of professional publications is the use of referees. A paper submitted to a journal is normally sent to two or three experts in the field for their comments. These referees, protected by anonymity, are expected to ensure that prior work is adequately acknowledged; that (in experimental descriptions) adequate precautions have been taken or appropriate corrections applied, or (for theoretical papers) that the mathematical techniques are appropriate and the calculations correct; that the data and their analysis are sufficient to warrant the conclusions which have been drawn and that alternative explanations have been considered; and that the paper is clearly enough written to permit other readers to assess the results for themselves. Clarity requires not only literary style but also sufficient mathematics in any complicated derivations, adequate and clear drawings and graphs and sufficient detail within the usual editorial need for brevity.

Papers are often returned to the authors with the comments of the referees. In many cases the suggested changes may be easily made and the paper resubmitted. Occasionally, though, a paper may be simply rejected, or the author may be unwilling to make the suggested changes. Reasons for rejection vary. Sometimes the work is of poor quality with cosmic deductions being based on miniscule evidence. Sometimes the author has chosen a journal inappropriate to his paper and may then get acceptance elsewhere. (For instance, there are journals that specialize in mathematical topics; others deal only with experimental methods and apparatus, and others with the pedagogical aspects of a science.) Sometimes, unfortunately, rejection of a paper may result simply from bias on the part of a referee. If it is thought that a referee is being unreasonable, an author may request additional reviews, and the editor (who cannot himself be expected to be an expert in all parts of his subject) will usually be strongly guided by the majority opinion. In the end an author can submit his rejected manuscript to another journal, often successfully. The appearance of a paper in a reputable journal is then more than merely the recording of what one more scientist thinks he has observed or calculated. It carries some of the weight of the journal's reputation and the knowledge that this paper has been carefully scrutinized. In most fields it is recognized that some journals are more selective than others and enjoy a higher esteem. Few cranks seem to be willing to submit their work to this reviewing system, or else they fail to meet the standards set. On the other hand, professional scientists accept and support the system implicitly, even though they may on occasion disagree with a referee.

This severe system of internal discipline has been one of the major factors in the growth and strength of science. No system can be expected to be perfect, and occasionally papers do appear which are at best questionable and at worst obvious nonsense. (Some years ago a very poor paper appeared, which acknowledged the advice of a friend of mine. When he was queried as to how he could possibly have helped with such a mediocre piece of work, his reply was that he had suggested that it not be published—and his advice was being acknowledged.) But, for the greater part, the refereeing system ensures that what is in print is probably correct and will not mislead those who may need to refer to it. The only reason that such a system is as successful as it is is that it is based on an agreed set of objective criteria, internally established, and operated by those who have chosen to abide by its results. There are standard ways of analyzing data, standard mathematical techniques, and known types of corrections to be included in calculations. It is rare that opinion enters. When it does, it may be in some more speculative field, where the paucity of data permits the postulating of different models that must be clearly described before being explored. Opinion may influence the conclusions being drawn but should not enter into the collection of data or their primary analysis.[3]

It is interesting to contrast this system with reviewing in the arts and literature where opinions may be strongly held and where fashions play a far greater role. (Fashions in science can take the form of emphasizing certain principles or techniques.) A scientific referee should at least pretend to be clothed in objectivity, but George Bernard Shaw as a music critic in London in the 1890's rejoiced in being partisan:[4]

> *People have pointed out evidences of personal feeling in my notices as if they were accusing me of a misdemeanor, not knowing that a criticism written without personal feeling is not worth reading.*

and[5]

> *No man sensitive enough to be worth his salt as a critic could for years wield a pen which, from the nature of his occupation, is scratching somebody's nerves at every stroke, without becoming conscious of how monstrously indefensible the superhuman attitude of impartiality is for him.*

So deeply ingrained is this form of criticism that it has spilled over into the controversy that was aroused by C. P. Snow's writings, obscuring the major points being raised and deflecting attention to secondary matters. In the humanities criticism of works produced long ago may gravitate toward a generally agreed assessment, but contemporary works—literary, musical, artistic—are notoriously prone to providing the inspiration for violently partisan feelings, which find expression in trenchant reviews and feuding factions. It is indeed a constant source of amusement to read the contemporary comments on first performances of works which are now accepted as great.

Contrast this with contemporary science: while there may be strongly opposed schools, there must still be agreement on the rules by which they must be judged. How much slower would be the progress of science if acceptance of a manuscript depended on the quirks of some irritable reviewer whose bias might vary from day to day, or even worse, who completely rejected certain types of papers. How often, by comparison, has one read a concert or theatre review and been left wondering whether one was at the same performance as the critic?

A very strong case can be made in support of the scientific reviewing system, but at the same time it must be recognized that it carries with it the possibility of censorship, whereby the unorthodox can be stifled. With the present multiplicity of journals, this is highly unlikely but not impossible. The unrecognized genius, rejected by conventional science for his unorthodoxy and with his papers denied publication, is probably nonexistent today. More usually such apparent cases involve the persistent refusal to consider an enormous body of scientific fact and instead the focussing on some obscure items selected for their support of a preconceived theory. Martin

Gardner in his *Fads and Fallacies in the Name of Science*,[6] has documented many such cases, which make interesting but sad reading. Although the scientific radical will encounter difficulty in having his works published in a professional journal unless he is prepared to meet the regular criteria via the reviewing process, there is still a way in which he may present ideas to a professional audience, normally in person, at professional meetings.

Although professional journals will refuse to publish papers that do not meet their reviewing standards, there are fewer restraints on other commercial publishers, and most university science departments receive a steady trickle of published works, announcing and denouncing, with sophistication and free imagination alternating as some new theory is propounded. Usually, these works attract little attention, but occasionally one is vigorously promoted. When this happens, the general public is often confused between science and pseudoscience, and the scientific community comes in for criticism because of its alleged attachment to dogma. Although such publications lie well outside the area of professional publication, they have been mentioned here because of the allegations that are usually made concerning the closed nature of the scientific journals. Underlying such allegations is a misunderstanding of the rules and operation of the journals.

To return to our main topic: The range of professional publications is enormous. Some are narrowly specialized, some quite general in their scope. Some are published by learned societies, such as the National Academy or the Royal Society or the American Chemical Society. Others come from commercial publishing houses but still with full editorial control in the hands of a board of scientists. Most journals appear monthly, some bimonthly or quarterly. Some, such as *Physical Review Letters*, appear weekly. The common factor to all of these journals, in addition to their use of referees, is their publication of original contributions. These publications are the primary sources in science and will be used repeatedly influencing later work. Their frequencies of appearance depend on the pressures of the field.

Each scientific paper, when published, will have recorded the date on which it was received, sometimes also with the date on which it was received in revised form. Priority of discovery can be attributed from these dates, but they also demonstrate the delay from receipt until actual appearance. These publication delays, which include the time taken for the reviewing process, may amount to many months, sometimes closer to a year. It has been argued that this is still usually a short time compared to the total time from the original idea, but to the scientist waiting to have his work published it sometimes seems unreasonably long.

Less formal methods have therefore been devised to hasten the flow of information. Within a narrowly defined subspecialty, there may be only a relatively small number of workers, who are mostly known to each other if not in person then at least by reputation. It has become customary when a manuscript is being submitted for publication or upon acceptance to send copies to one's friends or to those known to be interested. Such "preprints" are by now very well produced (given the excellence of Xerox and similar processes) and are a far cry from the poorly produced preprints of even ten years ago. To a very large extent, this distribution of preprints has replaced the personal correspondence of earlier days.

The system of preprints is not without its disadvantages. It fails to meet two of the most basic requirements of scientific literature: General accessibility, and scrutiny by referees. With mailing lists necessarily restricted by costs, there is the perpetual problem of availability to the readership of all who might be interested. Furthermore it is standard practice in scientific papers to include references to other papers with which the new results are being compared or on which they are based. How useful is it, though, to refer to a preprint that is not easily available? As for refereeing, although many laboratories and research workers are so well known that their papers will almost surely be published, critical review is still useful. How much weight then should be attached to the contents of a preprint that has not yet been reviewed? Clearly this depends on its authors and their prior reputation. Despite these drawbacks there are some fields, especially in the life sciences, in which the preprint practice is so widespread that experimental distribution centers have been set up to facilitate the broader availability of preprints. The scientific community is clearly much divided over this method of dealing with the problem of rapid distribution of information, and various alternatives have been discussed.[7]

So strongly is the need for speed felt in the field of high-energy nuclear physics and elementary particles that *Physical Review Letters* will publish unrefereed letters subject to rigid limitations on length and the requirement that "the subject must be new experimental results in high energy physics." In addition, a covering letter is required requesting this preferential treatment, which "must be signed by the department chairman or group leader under whose guidance the work was performed."[8] This is in marked distinction to the general requirement for review prior to publication, and it is too soon to know whether it will remain the exception, or whether it is the first of many exceptions, which will radically transform scientific communication if generally adopted.

Retrieval of published information represents another important current problem. It is often necessary to be able to have prompt access to the results of others. Such access is very greatly aided by the use of specialized journals devoted entirely to this task. By now most papers are headed by brief abstracts, written by the authors, in which the most important results and conclusions are listed. Compilations of these abstracts are published regularly in the separate areas of science with extensive indexes of authors and cross-listed subjects. Reference to appropriate abstracts allows one very quickly to assemble references to all papers on a particular subject, and those of marginal interest can be quickly discarded before one proceeds to the original journal. A glance at *Physics Abstracts* shows some of the dimensions of the problem. Over 1300 journals are now covered, and in 1969 there were slightly over 50,000 abstracts published. Of course, many of the journals are very specialized, perhaps with small circulations and few papers, but in this way their contents can become known much more widely, at least in summary form.

Almost all fields are suffering from an explosive growth in their professional literature. New journals spring up, to meet real or fancied needs. Some cover new disciplines, some are devoted to critical review articles, others attract short research notes, while yet others are restricted to purely technical matters. A survey of the library of any science department of a university will quickly reveal the extent of this headache. Committees and commissions come and go, making their recommendations toward a more rational organization of the scientific literature, its storage and retrieval of information, and the improvements that mechanization and computers will bring. It is probably fair to say that most working scientists, while admitting the existence of a problem, are skeptical of the proposed solutions or simply not interested. Most of us come to terms with the literature in our field, being highly selective of the journals we scan and the papers we read in detail, recognizing that we can cover probably 80 percent of the important papers with our present limited efforts. We could more than double our efforts, yet find few more papers of importance to us. Simply checking the papers cited by someone else in the field will usually help to locate some of those we have missed; for the rest, there is simply not enough time. We could make a profession of sitting in the library to the sacrifice of our own original efforts. Some duplication of work may result from this less than complete survey of the literature, but it is unlikely to occur in the very active fields at the frontiers of science and is probably not serious in other areas.[9]

Modern science dates back only to Galileo and Newton. When publishing their observations and theories in the seventeenth century, they wrote long and detailed treatises. With professional periodicals today being the almost exclusive means of reporting new results, scientific books play a different role. As textbooks they are used in university and college courses at various levels. Advanced texts can be used for graduate courses or for reference, and it is rare indeed for originality to appear between two hard covers. Today books represent consolidation and review rather than originality. Indeed, the prime requirement for a textbook is clarity in which current knowledge is gathered together, as opposed to an original paper where, of necessity, the emphasis must be on the novelty and brevity.

One more matter deserving attention here is that of language. Science is universal, and scientific writings are published in most countries, but the international language is now English. More and more, English has become the language of international conferences, and particularly in physics, many foreign journals are now mostly in English. So, for example, the Italian Physical Society's *Nuovo Cimento*, the Swedish *Arkiv fur Physik*, and the *Journal of the Physical Society of Japan* are almost all in English. This trend towards a wider use of English is coupled with the natural desire of scientists to have their papers read by the widest possible audience. Together these two factors pose a serious problem for many small countries. Is it better to have one's own journal published by the local national academy in the native tongue and containing all or most of the work done in that country, or is it preferable to publish the best papers in the major international journals and ensure prompt and ample recognition, but at the same time to leave only the pedestrian works to be reported in the local journals? There is no easy answer. In the field of international commercial aviation, the government of a small country may subsidize its national airline to ensure easier access to businessmen and tourists on whom the country depends. A similar scheme will not work in science; although many smaller journals do exist, their role in the progress of science is marginal, and most are undistinguished.

A special situation exists in regard to the Russian literature. Despite the surge of interest in learning Russian in the post-Sputnik years, it is in French and German that most scientists in the United States are best able to hobble along. With the large output of papers in Russian, translation services have grown up. The American Institute of Physics publishes bimonthly translations of the major Russian physics journals, and these are only a few months behind the dates of original publication.

As an indication of the distribution of languages, we can look at the results of an analysis[10] of over 27,000 abstracts in *Physics Abstracts* for 1964: Over 18,000 were in English, nearly 5000 in Russian, about 1600 in German, and 1400 in French.

How can we summarize this apparently chaotic situation with scientific information? It should be commented that to the insiders it does not seem as disorganized as it does to those outside. With the concentration of important literature and with the general awareness on the part of most scientists as to where to find their information, the system is probably more efficient than it has any right to be. But perhaps we should be thankful that our information dissemination and retrieval are not more efficient, or else we would be hard pressed to find time for our own work. If through lack of awareness, we should occasionally duplicate work done elsewhere, perhaps what we really have is a fortunate self-regulating mechanism that will prevent science from advancing too rapidly.

2.3 Personal Contacts

Correspondence between scientists has declined as air travel has improved, bringing with it the opportunity to attend conferences and meet one's professional colleagues and rivals. The growth of scientific conferences has been spectacular, and the exotic locations of major international conferences are the envy of our friends in the humanities. While force of circumstances may result in a meeting being held in Atlantic City or Miami, there are clearly good reasons for meeting instead in the Pyrenees, or in Kyoto, or in Leningrad, or in Hawaii, or in Sao Paulo. Is this travel worth it? Those of us who have attended some of these gatherings will usually (and honestly) say yes, very definitely. Conferences come and go depending for their interest on the state of the subject at any particular time; some conferences will be memorable for the many exciting advances reported or the cosmic surveys delivered with insight by an acknowledged master of the spoken word. Other meetings will be useful, but no more. But transcending all of these aspects are the personal contacts and the often deep friendships one makes and renews at intervals. With so many international tensions, these links may provide one way in which sincere international contacts can have very positive and widespread gains. In many countries leading scientists may have political influence, and a mutual respect can only help.

Clearly there are political aspects and possible implications to decisions to hold a conference in one country rather than in another, and these are discussed in Chap. 5. In the present section, we prefer to concentrate on the use and format of the conference itself.

Conferences take place at local, national, and international levels. Most national scientific societies hold several meetings per year, distributed around the country. Specialized or regional sections also hold meetings, and all of these are open to members of the society. The usual form consists of sessions of related papers that have been contributed by the members, each usually limited to a 10-minute presentation. For large gatherings, there are parallel sessions covering topics that seem not to overlap in interest. There are also some sessions devoted to a few papers given at the invitation of the organizers. These are longer papers, either reviews or original reports by those whose work is of great and general interest. Occasionally, a panel discussion occupies a session. For such gatherings, the program includes a compilation of abstracts, which guides one in the choice of sessions to attend. In all honesty, it must also be said that the most interesting part of many meetings is not the content of the sessions, where the level of presentation of papers is frequently abysmally low, but rather the informal discussions in the corridors.

International meetings tend to be different in at least two respects. The first difference lies in attendance which is usually determined by invitation. Many organizations try hard to keep attendance to manageable proportions. (Too large a meeting imposes a great burden on local facilities such as suitable meeting rooms and hotel accommodations, and personal contact tends to diminish as the meeting size grows.) Restricted attendance has been the cause of some complaints: Should preference be given to younger and often very active scientists, or to more senior men who may not themselves be as close to all current aspects but who still direct research projects? The second difference lies in the fashion of publishing conference proceedings. For normal meetings in this country, the papers presented usually find their way into the regular journals. Often the papers are progress reports on continuing or very recently completed work, but for international meetings there has been a steady drift away from this view of a meeting, to a position where the presentations are considered as finished products, which should be complete, even to having lists of references. At one time, the papers presented to an international conference were rapidly and simply printed, so that there should very soon be available an up-to-date summary of the state of a field but without the polishing expected for archival papers. Fashion has changed that. It is now, unfortunately, quite usual for a conference report to appear at any time up to six years after the meeting. Authors too often refuse to submit their manuscripts promptly, sometimes being simply dilatory, sometimes desperately adding more and more material to try to be as up to date as possible. Finally, a $40 volume appears which most science libraries feel compelled to buy. In no other area of human endeavor does last year's Edsel have such an assured market.

There is something, though, which is unique to attending a conference, and which cannot be supplied by the written word. To hear a man describe his own work is to add a unique dimension to one's picture of him. Some scientists are showmen; some mumble; one famous man uses his hands and expressions like a dancer from India, his affability and elegance of phrase increasing as the quantitative content of his talk decreases. Others are concise, even terse, but penetrating. After attending a conference and meeting a scientist, one cannot again read his papers in a journal without adding one's personal estimate of the author. It is the compound of these impressions that leads one to a very subjective evaluation of a scientist as being absolutely reliable or another as given to wild and poorly based flights of wishful fancy.

As already mentioned in the previous section, there has been a trend to the greater use of English at international meetings. Some meetings, however, designate two or three languages as official languages, and at some gatherings simultaneous or prompt translations can be heard with headphones. Knowledge of other languages is still of great use in fostering contacts, not only between conference participants but also with the local population.

Another avenue for personal contact needs to be mentioned, that of extended visits to other laboratories. As the work of a group or individual gains a reputation, students and visitors are attracted, some to stay only briefly, others for years. This is a very old tradition, now badly hampered by the shortage of funds. One can read a man's papers and listen to him at a conference, but working with him adds something. At first hand one can observe those small but numerous points that make the difference between competence and elegance, between thoroughness and insight. Science is the richer for the personal contacts, and their extension must be encouraged.

2.4   Science for a Wider Audience

The vast majority of the population are not scientists, yet they have contact with science and technology in many ways. Some have a general interest that may be satisfied by reading, while others are very actively engaged in hobbies. So, for example, the amateur astronomers are well organized into clubs and associations, and many of them help in the accumulation of valuable observational data on a worldwide basis. On a wider scale, public interest and awareness is sporadic, clearly dependent on the selectivity of the press, radio, and television. At present, it tends to center on sensational items: the transplant of a heart; a moon voyage; Nobel prizes. For the interest of the media and press to be aroused, there seems to be the need for one of two ingredients: a medical topic that affects millions, such as the relation between smoking and cancer, or the use of a medicine such as the contraceptive pill; or else the national security, as in decisions relating to the testing of nuclear weapons or the deployment of an ABM system. Regretably, steady progress does not seem to be newsworthy—it is the novelty or the alleged break-through that draws attention.

For those whose interest rises above the casual, there are excellent periodicals available. Over the years, the *Scientific American* has maintained a high level of popularizing with integrity. Its articles are written by active scientists; its illustrations are outstanding. Its range is wide, from archeology and anthropology, through quasars and pulsars and galaxies, to transit systems and mining. Besides articles that describe the latest scientific advances, there are others that deal with social impact. It is a truly great popularizer.

Quite different magazines are *Nature* and *Science*, both appearing weekly. Started in 1869 and now published by the Macmillan Company in London, *Nature* at one time was perhaps the main forum for announcing new discoveries, through its "Letters" column. Although other journals now attract many more papers, *Nature* still retains some of its older flavor and carries many important papers in radio astronomy and molecular biology. It has also, under its present editor, considerably increased its coverage of news in the area of science and public policy. *Science* is published by the American Association for the Advancement of Science (AAAS), but in other ways is very similar to *Nature*. Both publish brief letters, in which new results are reported. Somewhat longer articles are also included, but there tend to be only half a dozen per issue. Both devote regular sections to the reporting of scientific news, such as the progress of legislation affecting the funding of science, or actions that might affect the universities or national laboratories in which so much research takes place. Topical scientific news is covered at a semipopular level, and both magazines include editorials. (Editorial opinion is rare in a professional journal, apart from those occasions when the editor feels it necessary to berate his authors for their style or need for brevity. It is virtually unknown for a journal editor to express an opinion on any matter not very closely related to the management of his journal.) Depending on one's field, *Nature* and *Science* may be part of one's professional reading, but they also form very useful extensions by keeping their readers informed more broadly.

In a variety of other periodicals, science is presented to more or less specialized groups. Amateur astronomers are well served by *Sky and Telescope* and *Popular Astronomy*. These cater to the vigorous community of amateurs and contain excellent reviews of advances in astronomical knowledge as well as very practical advice on telescope construction and use. *Physics Today* goes to all members of the societies affiliated to the American Institute of Physics, and contains semipopular reviews, book reviews, news, and professional items, but no original papers. *Geotimes* (published by the American Geological Institute) and *Chemical and Engineering News* (American Chemical Society) serve somewhat similar roles for their respective communities. The Society of the Sigma Xi publishes the *American Scientist* bimonthly, and this includes reviews designed to appeal to nonspecialists. All of these journals, in varying degrees and at varying levels, help to disseminate news of scientific advances to a wider readership that is still, however, closely tied to the community of working scientists. Beyond this there is a growing number of periodicals that treat such areas as the history and philosophy of science.

None of the periodicals mentioned so far has a truly mass circulation, and so the science they describe and on which they comment is still far from most of the public. Few daily newspapers have regular science correspondents, and as a result only sporadic coverage is given to science and related topics. The *New York Times* is a conspicuous exception: It has a daily index to its major contents, and typically two or three of the items (out of about 80) are in the category "Health and Science." (Amusements usually amount to about 12 percent, business also 12 percent, and sports 15 percent.) The large-circulation weeklies, such as *Time* and *Newsweek*, have science columns that reach more readers than does any newspaper, but must obviously be highly selective with their limited space. While other weekly and monthly magazines also have allotments for science and most daily papers will have occasional items, the overall coverage is so scanty that the public remains poorly informed and/or uninterested. The extent to which this is true was revealed in a survey in 1957 which showed that about one quarter of those asked had never heard of fluoridation and nearly one-third had never heard of radioactivity. It would be interesting to know how a similar survey would come out today.

Faced with this sort of ignorance, we might ask whether more people should have at least some awareness of science and its applications. It can be strongly argued that this is more than desirable; it is necessary. Devising practical solutions is not easy. Simple and apparently obvious suggestions may easily overlook the complexity of the situation. For instance, just increasing the number of column inches devoted to science will not, in itself, suffice. It is hard to popularize science as it requires at least a minimum of intellectual effort or attention—scientific material when well-written requires reading habits different from those that most people are accustomed to exerting. What is needed, beyond this, is a combination of the skill of a good journalist and the knowledge of a scientist. This is easier to suggest than accomplish.

Attempts to improve the contact between the press and scientists have sometimes helped. Some professional groups have held workshops for science writers on topics of current interest. The National Association of Science Writers and the Council for the Advancement of Science Writing have helped. What is still needed is the cooperation of the individual scientists, with much more patience than some have displayed. After taking trouble to explain one's research clearly in an interview, it is annoying to find the published product unrecognizable, and many scientists have recoiled from further contacts with the press. Perseverance is needed and cooperation from the press, which can be best accomplished by assigning the same people regularly to cover science stories, enabling them thus to build up some expertise. Scientists need to avoid retreating into their laboratories, while newspaper editors need to feel more of an obligation to present their readers with material that is interesting even if not sensational. Radio and television could also take far more initiative than they have so far shown.

Even if massive efforts were to be made at presenting science to the public, it is probably unrealistic to expect that more than a small fraction of the public will take more than a casual interest, and in this respect science falls into the same category as an appreciation of poetry or the quartets of Beethoven. Relevance, though, does make a difference. Because of the uses to which science is put, and the way in which these affect major decisions on a national and international scale, it is no longer desirable to remain resigned to science being remote and its products understood by only a select few.

Recognition of this changed relation between science and society has come only in the years since World War II. The organization of the Manhattan Project for the production of the atomic bombs concentrated scientists on a single goal in a way that was unique. From their concern over the first use of these weapons, there emerged the Federation of American Scientists and the *Bulletin of the Atomic Scientists*, which still appears, published monthly. This periodical is devoted to sociopolitical discussion of the impact of science and technology on a worldwide basis. It does not publish original scientific papers, but is instead a valuable journal of review and comment. Over the years, it has chronicled the varying fortunes of American science, it has reported on science and its politics and problems in the Soviet Union, in China, and in developing countries, and it has provided an unique forum for the expression of a wide spectrum of views. A collection of some of its articles has been published under the title *Atomic Age*.[11]

The *New Scientist*, published in England, is a much newer journal that covers similar ground to the *Bulletin*, but tends to a slightly more popular level, with shorter articles and reports. In some ways, it runs parallel to *Science* in its reporting function, but without including original scientific communications.

Environment, published by the St. Louis Committee for Environmental Information, represents a completely different approach. Its aim is to provide an impartial documentation, at the popular level, on matters of major environmental interest. Such issues tend to become the subjects of strong and partisan debate; they include the environmental effects of power stations, the effects of pesticides, and problems arising from radioactive debris from nuclear explosions. Articles in this monthly magazine are, therefore, devoted to accounts of the scientific aspects, of current topics at a level that an interested nonexpert can understand. The articles are written by scientists as well as experienced staff writers, and are fully documented. The intention is to allow the readers to form their own opinions on the issues raised, but with the anticipated risks and benefits of the technology as clearly set out as current knowledge permits. Without such knowledge, it is hard to see how rational decisions can be reached. *Environment* differs from other periodicals in that it is not concerned with reporting advances in science unless they appear to have environmental consequences; it also differs from several other journals which are concerned with the sociological implications of applied science but do not usually discuss the science itself.

The public can become aware of scientific and technological problems in another way, that is through books. Especially with the growing market for inexpensive paper-backed books, national attention is occasionally attracted. Rachel Carson's *Silent Spring*[12] was one such book, and so, more recently, was Paul Ehrlich's *Population Bomb*.[13] When written by experts, such books can be major factors in triggering widespread and serious attention to some neglected problem.

While there are thus many ways in which scientific information can gradually reach the public, none is yet so efficient that we should feel satisfied. For improvements to occur, the scientific community will have to take a more active role and cannot afford to wait until it is called to account. Dramatic changes in the public attitude towards science cannot be expected overnight, but changes for the better will not occur without some effort on the part of scientists.

NOTES AND REFERENCES

1. J. M. Ziman, *Public Knowledge* (Cambridge: Cambridge University Press, 1968).

2. Thomas S. Kuhn, *The Structure of Scientific Revolutions* (Chicago: The University of Chicago Press, 1962).

3. See Harriet Zuckerman and Robert K. Merton, *Minerva* 9 (1971), 66 for a detailed study of the refereeing system, including analysis of the archives of the *Physical Review*, 1948–56. A shorter version of this article has appeared in *Physics Today*, July 1971, p. 28.

4. Eric Bentley, ed., *Shaw on Music* (Garden City, New York: Doubleday Anchor Books, Doubleday and Company, Inc., 1955), 34.

5. *Ibid.*, 58.

6. Martin Gardner, *Fads and Fallacies in the Name of Science* (New York: Dover Publications, Inc., 1957).

7. See, for example, *Physics Today*, **19** (1966), 38; and *Science*, **167** (1970), 1228, and **168** (1970), 194.

8. *Physical Review Letters*, **23** (1969), 629.

9. For a recent article on the use of computers for information retrieval, see J. T. Lynch and G. D. W. Smith, *Nature*, **230**, (1971), 153.

10. L. J. Anthony, H. East, and M. J. Slater, *Reports on Progress in Physics*, **32**, Part 2 (1969), 709.

11. Morton Grodzins and Eugene Rabinowitch, eds., *The Atomic Age* (New York: Simon and Schuster, 1965).

12. Rachel Carson, *Silent Spring* (Greenwich, Conn.: Fawcett Publications Inc., 1970).

13. Paul R. Ehrlich, *The Population Bomb* (New York: Ballantine Books, 1968).

certainty
and authority
in science

chapter **3**

It often seems to outsiders as though a scientist has undue confidence in the statements he makes, and that other scientists are too prone simply to accept such statements. Even when a scientist acknowledges a lack of precision, he seems paradoxically confident in defining his own imprecision.

In contrast, we are accustomed, for example, to a politician's statements being challenged. He may start by listing some facts, and then advocate some action or policy, designed to achieve some specified ends. It is not unusual for the "facts" to be scrutinized, by his opponents of the moment and by the public, and it may turn out that the "facts" were incomplete, or had been diluted by guesses, or were in some way more or less than the simple and unadorned data. Even when there is agreement on the facts, there may be great differences of opinion regarding other information that should (or should not) be included as a basis for the forming of policy. More generally, there are often very different sets of values employed in the formulation of policy. Finally, since political arguments take place in this country in English, they are at once open to participation by a wide public, unless the problems involve very technical matters such as details of economic policy.

The precarious nature of authority is characteristic of many areas in the humanities and social sciences. Rival schools of thought (or political parties) offer competing interpretations (or policies), and it is not unknown for personalities to be discussed in addition to theories or ideologies. As already mentioned in the preceding chapter, criticism that reflects such biases is recognized, for example, in literature, the theatre, and the arts. Scientific criticism is important, but usually takes a different form, and in the next chapter we shall look at some examples of controversy in science. For the present, I would like to focus attention on those aspects of the scientific method which permit scientists to seem so sure of themselves.

The precision of modern science would not be attainable without the development of what we call the scientific method. Ancient science was sometimes surprisingly accurate, as with the Babylonian astronomical observations, but even while their eclipse predictions were accurate, they were restricted by the often uncritical reporting of passive observations. Science in its modern form began to prosper in the seventeenth century, when it self-consciously started to undertake controlled experiments in which attention could be restricted to only a few variable quantities whose effects could be separately recognized and investigated. This procedure has permitted the exact specification of experimental conditions, so that an experiment can later be reproduced with the same results (within experimental errors). Reproducibility is crucial. With the vast range of science today, we must usually accept the results of others, but we know that, if we should have reason to doubt some result, we can in principle repeat that experiment. When repeated experiments fail to yield the same results, we must begin to suspect that other factors have entered but have not been recognized. Reproducibility is possible only when we know enough to be able to specify all of the important variables. That is usually easier in physics than in parts of biology, whereas it may be almost impossible in some areas of the social sciences.

Since reproducibility seems to be so important, just what do we mean by it? If I have three apples, it seems unlikely that I shall ever count them and arrive at 2, or 4, or 10. This is easy since we are involved only with integers (whole numbers) and are counting a small number. It is common experience, however, that we may count the number of people in a room and obtain 43 on the first count, 44 on the second count, and 43 on the third count. We would expect that the number of people actually present is a number amenable to exact specification, and we quite reasonably admit that we might have made an error in the second counting—perhaps someone moved and got counted twice. We would, in a case like this, tend to use 43 as the correct value since we got that value twice, or we might count again as a check. With care and possibly with repeated counting, we see no reason why we should not finally arrive at an exact answer without any error or uncertainty. We also recognize that this task becomes harder as the size of the crowd increases. For all practical purposes we cannot possibly expect to count a crowd at a baseball game exactly, at least not by sitting out in the middle of the field and using binoculars. For some purposes, however, we might be satisfied to know only the approximate size of the crowd, say to the nearest thousand. We could then make a careful count of a block of one-tenth of the seats, assume that all blocks are similarly occupied, and by proportion arrive at an estimate for the total.

What happens when we undertake a more complicated experiment, such as measuring the time taken for the swing of a pendulum? Here we keep the length of the pendulum and the size of the swing the same from one measurement to the next, and proceed to observe the times taken for a large number of single swings. In Fig. 1(*a*) we show the results of a compilation

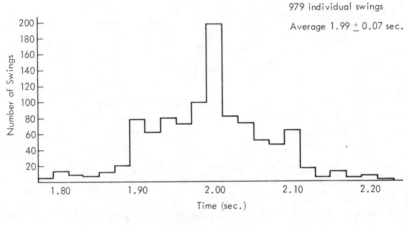

Figure 1(*a*)

of the data obtained during the course of this experiment in a freshman laboratory, when 979 pendulum swings were timed. The accuracy of reading the clock for each swing was 0.01 sec, and we have plotted in the graph the results of the measurements. (For convenience in presentation, the data have been grouped, so that those swings timed to have 1.90 sec and 1.91 sec have been plotted at 1.90; those for 1.92 sec and 1.93 sec plotted at 1.92, and so on.) It is clear that all readings did not turn out to be the same.

We notice at once that, despite the absence of perfect agreement, there is a marked tendency for the readings to cluster around the 2 sec mark, with about as many readings higher as lower. Also, the further we go from this central region, the fewer readings there were recorded. What we are observing in Fig. 1(*a*) is a "distribution" (of readings) whose shape turns out to be found in many situations where repeated measurements are made of a single quantity. The shape of the distribution can be mathematically described, and we can make many inferences after analyzing our data in this way. We can calculate the average, which turns out to be 1.99 sec. Figures 1(*b*) and (*c*) seem somewhat similar, in that they too show distributions that appear to be centered on about the same value as in (*a*), yet they are narrower.

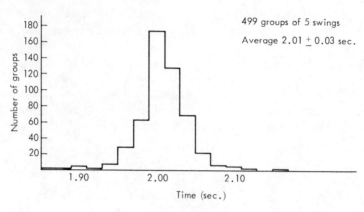

Figure 1(b)

The data displayed in Fig. 1(b) were obtained from the same pendulum as those for Fig. 1(a), but now the pendulum was timed through five swings together rather than individually, the measured time was divided by five to arrive at the average time per swing, and this result was plotted. Repeated measurements yield the distribution shown. For Fig. 1(c), the procedure involved timing groups of 25 swings, and plotting the average values so obtained.

Clearly, as we proceed to measure more swings and plot their averages, we improve our results, by which we mean that we have fewer readings that

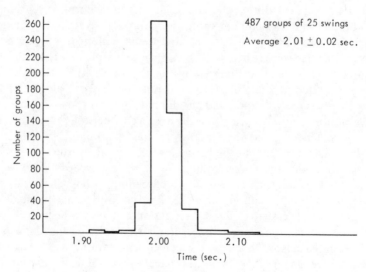

Figure 1(c)

lie far from the central peak. For each set of readings, we can calculate the average, and for this kind of distribution (known as the Gaussian or normal) the average will be on or very close to the central peak. The three distributions in Fig. 1 have very similar averages, but they obviously differ in being more or less spread out. We term the spread the "width" of the distribution, and it is a measure of the precision of our measurements; the more accurate our measurements, the narrower will be the resulting distribution. The width can be expressed mathematically from a more detailed analysis, and other mathematical properties of the distribution can also be defined.

Even without any mathematics, we can understand some of the reasons for the progressive change in going from (*a*) to (*b*) to (*c*) in Fig. 1. Each timing measurement involves some human errors in the starting and stopping of the clock. If we time only a single swing, then that one measurement has to carry the burden of any errors made, as for instance in judgment of the beginning and end of a swing. But when we time five swings by switching on once and switching off once, then the errors are shared between five swings when we calculate the average time per swing; similarly, when we time a group of 25 swings, we reduce the contribution of error to each swing. Although the detailed mathematical theory is rather more complicated than this simple outline, it should at least appear reasonable for the accuracy to improve when we measure more swings, and we see the confirmation in the narrower distributions.

The science of statistics covers these and many other topics, and has been extended through powerful mathematical methods. The interested reader will find some references in the bibliography; here we shall confine ourselves to some of the simpler inferences that can be drawn from the measurements we have been describing.

A detailed mathematical analysis of such a Gaussian distribution as we have been describing shows that 50 percent of the readings will usually fall within a certain specified distance on either side of the average; two-thirds will fall within a slightly larger but exactly calculable distance, while 90 percent of the readings will lie within a larger distance again. We are able to say that, usually, only one reading in a thousand will lie further from the average than a specified amount, while the chance of a reading being even a larger distance away is one in a million. In fact, for any range of possible readings, we can specify how many readings we would expect to obtain within that range. We can therefore summarize the data in Fig. 1(*a*) by saying that the pendulum swing takes on the average 1.99 sec and that an interval of ± 0.07 sec around the average describes the width of the distribution. This interval (0.07 sec) is termed the standard deviation of the distribution, and by this we mean that nearly two-thirds of all the readings fell within the range 1.92 sec ($= 1.99 - 0.07$) and 2.06 sec ($= 1.99 + 0.07$), while close to 95 percent of the readings fell within the range 1.85 sec

($= 1.99 - 2 \times 0.07$) and 2.13 sec ($= 1.99 + 2 \times 0.07$). Similarly, the distribution in Fig. 1(*b*) is described by quoting 2.01 ± 0.03 sec, while for Fig. 1(*c*) we obtain 2.01 ± 0.02 sec.

The calculation of the standard deviation of a distribution from the actual readings is a straightforward and routine task, and from it we can at once tell how closely the measured values group about the average. This is a reflection of the accuracy of the measurements.

Until now, we have been concerned with the calculation of the average of our measurements and with the way in which the individual values were distributed about that average. We could proceed much further with our analysis along standard lines, but that would make the presentation here unnecessarily complicated. We shall just note though that within the ranges defined by the various $\pm$ values the three distributions have average values in good agreement with one another, and we can see that the decreased widths of the graphs are reflected in the smaller $\pm$ values.

This method of analysis is universal. When a scientist states his results in this way, we know at once what to expect if we were to try to reproduce his data. Of course, there are complicating factors that require more sophisticated methods, but underneath there is a common language in the method of analysis.

Statistics goes far beyond these simple cases, and many other shapes of distribution are encountered when we record the results of our observations, whether of natural phenomena or contrived situations. It is quite surprising how amenable the most unlikely data are to statistical analysis. In his standard work on statistical methods, R. A. Fisher[1] refers to the data accumulated by Bortkewitch. Records were kept of the number of cavalrymen killed per year by horse kicks. Over a 20-year period, and covering ten army corps, there were 122 deaths, leading to an average of 0.6 death per year per corps. In some years there were no deaths, while at the other end of the spectrum one corps experienced four deaths in one year. Bortkewitch found that the number of times that the values 0, 1, 2, 3 and 4 deaths were recorded followed very accurately the distribution that is well-known to statisticians as the Poisson distribution. This distribution, named after a French mathematician, applies to rare events that occur quite independently of one another. Rare events that are related (i.e., occurrence of one influences the chance of occurrence of another) require description by a different formula. It is common experience that accidents seem to occur in groups, and statistical theory provides an explanation.

Statistical methods have been evolved to test a collection of data: We can test whether our data follow one distribution or another, and we can also say quantitatively how well they agree; we can give the likelihood of finding certain results in the future if we know the shape of distribution that applies. If we suspect that some occurrence (smoking) might be associated with some effect (cancer), we can use statistical tests to probe the level of significance of the suggested association. (We should also recognize that associations do not necessarily imply cause and effect, since other factors may also enter.) In making statistical tests, we need to assume the reliability of the data we are handling. Sometimes a test will reveal the presence of unsuspected sources of bias in the collecting of data or will confirm suspected sources. At other times, the data might be insufficient to permit any strong inferences to be made, and carefully applied tests will show this. We can then calculate just how much more data we need in order to make a statement at some desired level of confidence.

Consequently, when scientists analyze their data and compare them to those already published, there are well-established methods to be used. Good experimental design will include some consideration of anticipated accuracy and a calculation of how many measurements are needed, in order to be able to establish agreement or difference to the degree desired.

No measurements are absolutely accurate. If we time the pendulum swings to the nearest tenth of a second, the distribution of measured values will have a certain width. We shall obtain a narrower distribution if we use a better clock and read to the nearest hundredth of a second, but no matter how accurate our clock we shall always end up with a distribution of values. In the end, we summarize our experiment by quoting the average value and a range of values that tells us about the distribution. (It is customary to quote the standard deviation, but in some circumstances equivalent figures may be given, which are not quite the same and are termed confidence limits.)

We can now consider the following: We obtained an average value of 1.99 sec when we measured single swings of the pendulum, and the corresponding standard deviation was 0.07 sec. If one more swing is to be measured, what value will we get? We can say that 1.99 sec is the most probable value, but that with a probability of about 68 percent (i.e., two-thirds) we expect the next reading to lie within the range 1.92 ($= 1.99 - 0.07$) and 2.06 ($= 1.99 + 0.07$) sec. If we go ahead and make 100 further measurements, then we expect about 68 to lie within that range, and the other 32 outside. We cannot predict *exactly* what the outcome will be for a particular measurement, but we can set the odds, and we can be quite precise in our lack of precision. This kind of statistical prediction is of very great value, despite what may at first seem great vagueness.

The use of mathematics in science goes much further, and involves many branches of mathematics besides statistics. We find that we can represent and describe the relationship between many natural quantities in mathematical terms, perhaps as a summary of an experiment. Standard mathematical manipulations can then lead to predictions that can be tested by further experiment. There are deep questions involved in probing into the use of mathematics in science, and all is by no means yet understood. Nevertheless, as a language of science, mathematics has been remarkably successful; it is universal, and it provides a means of arriving at agreement amongst scientists as to their experimental results and theories.

There are still, however, places where opinion can enter, influencing a calculation or analysis, and some will be covered later in this chapter.

With the statistical method that we have been describing, we are in a position to explore the way in which a scientific "law" can be deduced for a simple system, namely our pendulum. All of the data listed so far were obtained with the pendulum length set at 100 cm. Suppose now that we change the length to 110 cm, and time the swings. We then obtain 2.10 sec. Changing the length again, to 120 cm, and repeating the experiment, we find 2.20 sec. The results can be very clearly displayed in a graph, as in Fig. 2(*a*).

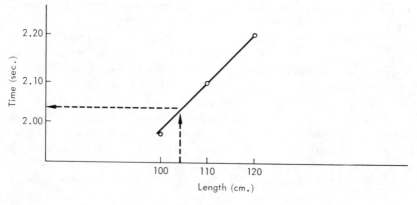

Figure 2(*a*)

We might wish to see whether there is some relation between the pendulum length and the time for a swing, and a straight line seems to go close to or through our data points quite nicely. We have made measurements only for the lengths 100, 110, and 120 cm, but it would seem perfectly reasonable to draw the solid line on the graph, and we could then read off the time value we would expect for a pendulum length of 104 cm, for example. This is the procedure of interpolation; we have not as yet actually made a measurement for the 104-cm length, but we are assuming that whatever physical regularity held for the measured points also holds for all lengths and times in between.

A somewhat different procedure is involved in going beyond the range of our original measurements, i.e., in extrapolating. The dotted lines in Fig. 2(*b*) show a simple extrapolation, for lengths less than 100 cm and more than 120 cm. We assume that the straight line connecting our three points can be extended as shown. From the graph, we can then read off a time of 1.43 sec to be expected for a pendulum of length 50 cm, and a time of 3.56

sec for a length of 250 cm. We proceed to make the actual measurements at these lengths and get 1.42 sec and 3.19 sec, respectively. How do we reconcile these values with those we have just predicted on the basis of our rather confident analysis and extrapolation?

We might try to shrug off the measurements as being possibly chance results, and we might hope that when we repeat them we would get values closer to those expected, although the agreement seems already good for the data for 50 cm. Perhaps we were careless and miscounted the numbers of swings, or misread the clock. Sometimes it is very hard to locate an experimental error other than by many repetitions that agree among themselves, leaving the original one isolated. Perhaps, though, we might feel greater confidence in our experimental methods, and in that case we can take the hint that perhaps we are dealing with a more complicated situation than can be represented by a straight line.

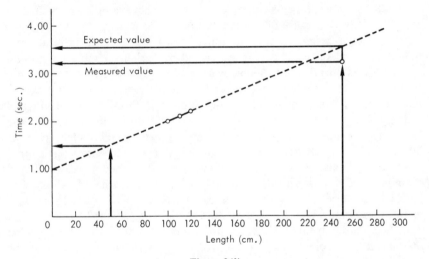

Figure 2(*b*)

Using our standard statistical tests, we can soon satisfy ourselves that our results for a length of 250 cm are (with a very high degree of confidence) *not* on the dotted line that was extrapolated from the measurements in the 100–120-cm range. At this point we retreat, and try to find a more complicated expression for the relation between the length and the time for a swing. The result of such a calculation ultimately shows that the time is proportional to the square root of the pendulum length, over the ranges of lengths we have examined. So far, so good—we have a somewhat better experimental description. We can even call this an experimentally derived law, and use it for further predictions. The interest is greatly extended when we can, from quite independent assumptions, produce a theory that has this relationship as its result. We can, in fact, construct such a theory if we start with a highly idealized description of a pendulum (we call such an idealized representation a "model"): We consider the pendulum to consist of a small rigid object at the end of a cord which has itself negligible weight and which is attached at the top end to a rigid support. We assume that gravity is the only force acting. We apply one of Newton's laws of motion (obtained or postulated separately), and can predict that the time for a swing should behave in just the way we have found by experiment.

By now we have quite a lot of confidence in our understanding of this relatively simple system and can proceed to add complications, investigating the effects at each stage: What happens if we take into account the air resistance felt by the swinging pendulum; how about allowing for the weight of the supporting cord? This approach, applied to more and more complicated systems (or models), has served science very well, and we have assembled many "laws" in the process by which we can predict the outcome of further experiments or the course of various natural phenomena.

We should recognize that scientific "laws" are neither arbitrary nor immutable. Unlike the laws that are passed by parliaments and congresses, scientific laws do not reflect ideologies, election promises, or political compromise; rather, they represent the codification of observations and experimental results, sometimes majestically simplified through the vision and insight of a Newton or an Einstein. Coming from human thought and following experiments whose accuracy may be great but still limited, scientific laws are subject to change and extension. When new results are reported that seem to challenge the validity of some well-established law, we should examine these results critically. Do they lie within the range of validity that has been established for the law in question? If within the range, what new features are there which seem to contradict the older results? Have the new data been carefully analyzed, and do the conclusions flow logically and unambiguously from the results? Furthermore, what do we know of the scientist who has produced these results? An established scientist might leave some precautions unmentioned, his own experience and reputation vouching for adequate care having been taken. On the other hand, some scientists are known for their less-than-careful work and for their haste in claiming new results, which later fail to find confirmation.

The subject of the nature, validation, and breaking of scientific laws has received much attention from philosophers of science, and the works of Kuhn[2] and Toulmin,[3] in particular, can be consulted as starting points. Although this subject is an active branch of philosophy of science, it is interesting to note that most scientists are not, in fact, guided by its study. What the active scientist does notice, however, is the reliability of his colleagues' work, as judged by himself and others according to their professional criteria, such as reproducibility, statistical accuracy, and consistency with other experimental results.

When we add up the points we have been describing, we approach the recognition of scientific authority. There are generally-agreed-upon methods for evaluation of data; there is a standard language (mathematics); and there is the role of refereeing of papers that will appear in open journals. Scientists establish their reputations within the system by having their results and theories confirmed, directly or indirectly, through the work of others.

An interesting example of differing views of scientific authority occurred a few years ago, in the controversy surrounding the theses propounded by Immanuel Velikovsky.[4] He had assembled an impressive collection of ancient legends and from these had suggested a radical view of the development of some parts of the solar system, the planets Venus and Jupiter in particular. His theories presented a direct challenge to the widely accepted theories and laws of mechanics, gravity, and electromagnetism, and they came under strong attack from many scientists. First widely published in 1950, these views have led to a controversy that has erupted intermittently and still attracts much comment. The interesting point for us, at this moment, is that his books were reviewed, in a variety of journals, by many eminent scientists, who heaped scorn and ridicule on them. Most other scientists accepted the criticism, often without reading the books for themselves or going through any calculations or other checks for confirmation. The authority of the scientific reviewers was accepted, whereas the laudatory reviews by nonscientists were ignored or rejected by scientists. By contrast, many nonscientists, reading the books and their favorable reviews in nonscientific magazines, strongly denounced the scientists for so willingly accepting the word of other scientists instead of reading and testing for themselves. What was involved, very deeply but apparently not explicitly on the part of the combatants, was a lack of understanding of the nature of authoritative opinion in science (quite apart from a demonstrable lack of understanding of basic principles of physics, too). To many of the nonscientists, the opinion of one reviewer was quite as good as that of any other, and no distinction was drawn between those with experience in the field and others who were not known in the scientific community. To the scientists there was no problem: The critical reviews by scientists made it clear that they need not read the originals unless out of general interest, while the other reviews seemed often uncritical or tangential. (Some other aspects of this episode are discussed in Chap. 4.)

Earlier, mention was made that opinion could enter. When one wishes to make predictions beyond a region that has been experimentally tested, how should one proceed? There are many courses open. The simplest approach is to assume that the existing formula can also be applied in this new region (in terms of our earlier example, we could try to apply the formula even to a pendulum one mile long). Would this yield accurate predictions? Only experiment can tell. One might suspect, though, that other effects might enter (for a one-mile pendulum, the weight of the cord would seem to be of importance), and an attempt could be made to modify the original formula. It is in making these modifications that opinions and experience and intuition enter, in a very subjective way. When the system under discussion is amenable to direct test, then it is (in principle) an easy matter to distinguish between the various predictions and thereby gain a better understanding of the system's behavior under a wider range of conditions.

When scientific opinions persist in differing on matters such as the role of DNA, or the history of the solar system, then we may expect that the resolution of these differences will come from further experimenting and the application of the well-known methods. Pending successful solution, scientists will often speculate on the possible or probable final results and will use their speculations as guides in their work. The prestige of some scientists is sufficient that many others may well be persuaded into or out of lines of attack by such views. The situation is very different, though, when the problems are not entirely scientific or technical, but involve nonscientific factors, often political or economic.

A good recent example is that of the antiballistic missile (ABM). Here is a complicated system that requires the coordinated and successful operation of rockets, radar, computers, and explosives. Opinions have differed, even among highly qualified scientists and engineers, as to whether the entire system could operate with certainty, and opinions differed even further on the question of the desirability of constructing the system. Where does scientific authority stand in such a dispute, and what will the general public think when it sees scientists disagreeing?

It is essential in the discussion of such issues that a line be drawn with great precision between the science and nonscience. The scientific-technical aspects should hopefully cover a large area of general agreement as, for instance, the range and accuracy of the rockets proposed and the quantity of radioactive debris produced by a nuclear warhead. Then there will be a grey region, where data are simply insufficient for dogmatic statements: Has a radar of the needed complexity shown itself free of various suggested problems? From these two areas we pass to the political and strategic considerations: Is the ABM needed? What are the alternatives? What are the risks of proceeding with the installation of this system, and what are the suggested benefits?

Scientific authority can cover the first of these three areas; it should be used to demarcate the extent of the second; and by elimination, it sets the boundaries for the third where opinions, values and prejudices, and hunches and suspicions merge together in arriving at a political judgment.

We should recognize that scientific authority can be very strong, but that its domain is limited. Within that region, there can be good reason to accept the view of a scientist. When we enter wider discussions, many voices will be heard with equal authority. Within the strictly scientific area, scientific authority can be effectively challenged only by those who speak the language and are familiar with the problems and methods and accumulated successes. Challenges to the established scientific order will not succeed when inconvenient facts are ignored. Polemics and debating tricks have never yet overturned a scientific theory, although they do help to win elections.

NOTES AND REFERENCES

1. Ronald A. Fisher, *Statistical Methods for Research Workers* (Edinburgh and London: Oliver and Boyd, 13th ed., 1963), 55.

2. Thomas S. Kuhn, *The Structure of Scientific Revolutions* (Chicago: The University of Chicago Press, 1962).

3. Stephen Toulmin, *The Philosophy of Science* (New York: Harper Torchbooks, Harper and Row, 1960).

4. Immanuel Velikovsky, *Worlds in Collision* (New York: Dell Publishing Co., 1967).

controversy:
the resolution of
scientific conflicts

chapter 4

After all that has been said in the last chapter on the subject of scientific certainty and the criteria on which it is based, one might be tempted to think that science marches inexorably on, its successes accumulating as each new fact spurs on the next theory and an admiring scientific public breathlessly awaits the latest journals with their reports of discoveries and ideas. Science is a human enterprise, however. Despite the existence of well-recognized guide lines for judging new results and theories, some very human passions manage to creep in at times, particularly when prestige is at stake or when a new idea deeply offends some entrenched views. There can then emerge controversies that illuminate a very wide range of human emotions. Polanyi, in a perceptive essay, has drawn attention to some historical examples.[1] In the present chapter, one historical example and some more recent events will be used to illustrate different forms in which controversies arise.

Closely related to conflict is competition. This is a relatively recent development in science, in which experimenters are working independently toward a single goal whose significance and importance is generally recognized. Occurring at the very frontiers of scientific knowledge, such competition can be greatly stimulating to those engaged in the work. An excellent example is the race to gain a clear understanding of DNA, that complicated molecule so important in genetic processes. Watson has described this episode in his book,[2] which attracted much attention because of the way in which he described the personalities and progress. It is, indeed, these human aspects that provided so much general interest, for very few who read the book could have had any knowledge of biochemistry or of the use of X rays in probing molecular structure.

Similar rivalry and competition occur in the areas of elementary particles and modern astrophysics, for example, but often they are hidden from the public by a parochial jargon. However, major discoveries are now more frequently attended by greater publicity (often contrived or abetted by a university's astute public relations office), especially in medicine with subjects such as the genetic code or the physical basis of life. There is still too often a barrier against widespread interest in even such fascinating accounts as Watson's. Whereas science once dealt with the obvious and the apparent on an everyday scale, much of today's science deals with entities whose existence must be inferred (such as atoms) or concepts (such as energy) that are exactly formulated by scientists but inexactly understood, if at all, at large. This can be overcome, but it usually requires an accumulating background of information and also more intellectual effort than most of the public seem willing to expend or the media to foster. There is no easy way to grapple with the nonintuitive, and as a result many of the more interesting scientific controversies remain known only to the experts.

Scientists may also differ from one another in their interpretations of existing data. In new and vigorous areas, as the experimental results accumulate, there will be many attempts to codify them into a more or less comprehensive theory, and there will usually be several such attempts. These approaches to an explanation will differ in the weight they attach to the data at hand, and in predictions they make for further experiments. Kuhn[3] has discussed in detail this stage in the development of scientific ideas. In particular, he has pointed to the error in the widely held view that a theory is rejected when even one discordant fact is turned up; Kuhn correctly points out that a theory that has had a measure of success is often retained as a guide to further experiments, until a substitute theory is proposed which is clearly better. It should be added that this process which Kuhn discusses is effective when it takes place in the professional literature, with its usual requirements for the reviewing of manuscripts. Competing theories are put before the appropriate scientific groups through regular journals, and do not receive their main airing in the daily press. Conversely, scientists will view with considerable reserve, if not scepticism, theories that are propounded only in unrefereed publications. It is a measure of genuine scientific controversies today that they conform to the normal rules of scientific communication.

4.1   Opinions on the Nature of Light

As a first example of disagreement between scientists, it is interesting to cite a historical case that arose from differing interpretations of the known experimental facts. Newton and Huygens appealed to different models (or idealized representations) to describe light's properties. Reflection and refraction and the properties of simple lens systems were already known, and so were some of the phenomena of interference in which light passing through very thin films (as of oil) gives colored effects that change when viewed from different angles. Newton's *Optick's*, published in English rather than the scholarly Latin and containing virtually no mathematics (in strong contrast to his *Principia*, which is heavily loaded with geometrical proofs), was widely read and widely quoted by literary writers of the eighteenth century. By contrast, Huygens's theory did not gain widespread scientific support until over a hundred years later, when Thomas Young developed a wave theory of light far beyond anything Huygens had considered. (Young, incidentally, also made major contributions toward the deciphering of the Rosetta Stone.) What we see, in the case of Newton and Huygens, is an agreed body of data and facts that lent themselves to interpretation through more than one theory. The "truth" is not always easily established, however, and our views regarding the nature of light have oscillated over the years, as more and different effects have been observed.

4.2   Is There Agreement on the "Facts"?

Sometimes the disagreements between scientists arise over contested "facts." There may appear in the scientific literature a report in which a new effect is observed, but other experimenters may have great difficulty in reproducing the results. With complicated and sophisticated experiments, it is sometimes impossible to reconcile the different results or to understand which is right or wrong. An example of such a problem, still unresolved at the time of this writing, has to do with the announced discovery of a new form of water. For a long time, it has been known that groups of atoms can combine to form what we call "molecules": For instance, two hydrogen atoms together with one oxygen atom form a molecule of water. The properties of water are quite different from that of hydrogen or oxygen. Similarly, groups of molecules can sometimes be persuaded to link together, forming very large molecules, and just such a process is the basis of the modern plastics industry. Plastics derive their remarkable properties from the particular molecules that they comprise and the manner in which they are bound together. Quite recently, there have been reports that "polywater" had been discovered; that is, some experimenters have produced a liquid that contains hydrogen and oxygen atoms in the ratio 2:1, but with the liquid having other properties strikingly different from regular water. Despite considerable experimental activity, there is currently not even agreement on the "facts," and some scientists have suggested that the results to date are no more than the result of minute impurities which might be introduced inadvertently at various stages of the experiments. But, once again, there are recognized channels in which the debate is conducted, and the final facts will have to meet the normal criteria for critical and professional scrutiny.[4]

This last point should be emphasized: Until now, in this chapter, we have been considering scientific disagreements that are debated through professional channels. From these debates, clearer pictures emerge and science makes progress. It is the intricacies of this process which have attracted the attention of philosophers and historians of science and have been so well analyzed by Kuhn. There is another aspect to scientific controversy, however, which has not received nearly as much critical attention, and the remainder of this chapter will be aimed in that direction.

4.3   Disputes Involving More Than Science

The truly heated and protracted battles which fully justify the use of the term "controversy" are often fought around the edges of science, where a larger public has become aware of issues, and where nonscientists become involved. When religious or political views enter, perhaps because they are challenged by implications or misunderstanding of the scientific results, one can expect the clash to be strongest. Perhaps the greatest passions are aroused when non-scientists persist with an interpretation or theory or challenge which (in the eyes of the scientists) has at best a questionable basis. This type of controversy very soon changes into an attack on scientists for their attachment to dogma and their unwillingness to accept anything unless produced by a member of a closed professional club. Such arguments rarely, if ever, contribute to the progress of science itself.

These controversies, which are peripheral to the mainstream of science, differ greatly one from the other. Sometimes a foolish misinterpretation of a scientific theory (such as relativity) receives wide publicity, much to the resentment of the scientific community. At other times, the scientists' wrath is aroused by the widespread but quite uncritical public acceptance of poorly documented observations, which too often are considered still a part of genuine science. (Although an examination of such episodes can be entertaining, the necessary detail would take us too far from the thread we have been following. In pursuing these subjects, an elementary grounding in the appropriate science needs to be assumed or developed to illuminate the problems.) Too often, such discussions center on trivial points and are conducted as in a debate, whereas a few well-organized scientific facts will dispose of the question.

One scientific theory that has come in for more than its fair share of misunderstanding has been Einstein's theory of relativity.[5] Without going into any of its mathematical content, we can understand why it should have attracted so much attention and why there is still a steady trickle of books coming onto the market, challenging Einstein and his theories and urging the reconsideration of relativity, to the end of restructuring physics without using relativity as we now know it.

As experience accumulates, we call it intuition, and we use it to judge new theories and ideas. An inability to imagine within a newly suggested framework can stand in the way of understanding a new idea and can lead to controversies. This was especially the case with the theory of relativity. Einstein's use of the time dimension was radical and has still not been understood by many. In most areas of science, our observations are of slowly moving objects. The time taken by light to travel from these objects to us, thereby permitting the observations to be made, is so short as to be negligible. However, when one deals with objects or with radiation (such as light or radio waves) that are travelling with speeds not grossly less than the speed of light itself, then the light's own travel time becomes important. Since our everyday experience does not usually include objects moving with the speed of light, it is understandable that the apparently strange results that Einstein deduced should have been viewed with such surprise. This indeed caught the popular imagination. A second factor that surely contributed to the publicized myth was the alleged mathematical complexity of the theory—it was said to be so difficult, in fact, that barely ten men in the world could understand it. This has been much overemphasized: the special theory, introduced in 1905, did not in fact require abstruse mathematics, although the general theory, proposed about ten years later, is indeed more complicated but today is included in a fairly standard graduate physics course. A third reason for the public attention was the interest in Einstein himself, who became a stereotype for the figure of the theoretical scientist. Together, these various factors created an image of the theory and its originator. Today, the special theory is accepted as "correct," in the sense that its necessary inclusion in calculations leads to excellent agreement with experiments. Whereas some opposition in the early years came from within the scientific community, largely through conservatism or misunderstanding, the rearguard and guerrilla actions today do not come from established scientists, but rather from interested amateurs, and survive only by a strong-willed concentration on alleged deficiencies to the exclusion of an enormous body of other facts.

Unidentified flying objects (UFO's) are harder to classify. While arousing great public interest, they have not so far caused the scientific world to become seriously concerned, still less strongly partisan. The exhaustive study, directed by E. U. Condon and sponsored by the United States Air Force,[6] failed to turn up definitive evidence for the existence of any new physical pheneomena, nor did it substantiate the various claims involving extraterrestrial beings. Many of the sightings could be explained in terms of complicated but well understood physical effects. Some, in the end, could not be explained, but, as the authors of the report point out, that does not mean that new effects must be invoked. (We may remark here that natural pheneomena may be extremely complicated and thus very hard to disentangle into their constituent processes, but this is still far from requiring us to invent radically new theories.) What lent especial interest to the UFO reports was their geographically widespread origin, their association with the possibility of breaching the country's defenses, and even the possibility that we were experiencing visits by beings from some other planet. It is fair to say that there are a few scientists who are not satisfied with the findings of the Condon Report, and who feel that substantial further study should be pursued. But, as the report points out, nothing has been added to our scientific knowledge from over twenty years' study of UFO's. The public fascination remains, however, despite scientific investigations, and UFO clubs and magazines prosper, while books continue to appear in their support. It would seem as though UFO's fulfill some deeply felt need, but this is surely beyond the area of science we need to consider here.

Scientists became involved in the UFO affair when their expert advice was sought in response to the reports that began to appear in the late 1940's. The failure of the scientists to produce supporting evidence, or even, in most cases, to take much interest, did provoke criticism and there have been allegations of deliberate attempts, at various times, to stifle any truly probing study. But, as mentioned, the scientific community was largely unaffected, and the lingering interest lies elsewhere.

4.4   The Velikovsky Affair

In another affair that started in 1950, some scientists played a far more active role, much stronger views were expressed, and even 20 years later the embers of this controversy show intermittent signs of fire. The squabbles over the theories of Immanuel Velikovsky (mentioned in Chap. 3) soon departed from the relatively simple task of pointing to their quantitative shortcomings, to the trading back and forth of charges of bias, in which the actual science was too soon obscured.

Briefly, Velikovsky studied legends and myths in many cultures and came to feel that there were common themes which could best be explained in terms of natural catastrophes that had been observed in ancient times.[7] He suggested that a consistent understanding of the old legends could be gained if they were considered as recording the observed motion of a comet that had approached extremely close to the Earth, disturbing the Earth's rotation. According to Velikovsky's interpretation of the legends, the comet

had then receded and reapproached, and finally receded from the Earth
to take up permanent residence in an orbit about the Sun; this object (he
suggested) is now known as the planet Venus. While Velikovsky's books
make interesting reading, they do not stand up well under critical scrutiny.
Scientists who reviewed his books pointed to the shortcomings in his theory
and also to his evaluation and selection of evidence. It was also pointed out
that his theory could be accepted only at the cost of rather arbitrarily dis-
pensing with an enormous amount of scientific theory that had been highly
successful until then, in order to explain observations or interpretations whose
validity was, to put it mildly, not uncontested. Some of the reviews were
quite vigorous. Most scientists were prepared to accept the critical comments
of the reviewers. This is where the role of scientific authority is very well
illustrated, and it needs emphasizing.

When an eminent geophysicist reviews a book and gives his expert
opinion that it has little scientific value, this will generally be accepted by
other scientists. Supporting this apparently docile acceptance of authority
is the system whose operation has been described in the previous chapter.
There are standard ways of testing theories, there are many facts which a
successful theory is required to explain, and there is, just as important, the
reputation of the scientist who is making the judgment. This was, apparently,
not well understood by the many nonscientists who voiced strong objection
to the reviews and who asked why the views of any scientist should be so
meekly taken. What was indeed remarkable was the unanimity of the scientific
reviews.

No further notice would probably have been taken of Velikovsky's ideas
if the controversy had been confined to their scientific merits. But right from
the start, other matters intruded. At no stage have Velikovsky's views been
published through regular professional literature and have thus not been
subjected to the usual reviewing procedures. Some of his critics were indignant
that when his first book was published in 1950 it was widely advertised as a
science book. More than that, sensational reviews appeared in *The Reader's
Digest* and *Harper's* in advance of the appearance of the book itself. This
method of publicizing a new theory annoyed many scientists, and some
brought extreme pressure on his publisher. Under alleged threat of a boycott
of all textbooks from that publishing house, publication was transferred to
another publisher (who did not have a textbook division). An editor of the
original publisher and a planetarium director both lost their jobs allegedly
because of their support for Velikovsky and his views. The scientific community came under strong attack then, on several grounds. First was the
pressure exerted on the publisher; most scientists, though, will not even try
to defend such a tactic. But the attack was broadened, and the scientists were
indicted for their refusal to listen to unorthodox views, for their willingness
to accept the judgment of others, and for their prejudice in not being prepared

to consider ideas that did not come from an insider. (Velikovsky's training was in medicine and psychoanalysis.) The controversy abated from its early violence, but erupted again in 1963 and 1967.[8] Science has remained unshaken and Velikovsky's theories have not taken hold, not because he is outside the professional fraternity, but simply because these theories did not stand up to rigorous review. On the other hand, the nonscientific aspects would appear not to have been settled. What the controversy *did* illuminate was the lack of understanding of the scientific process on the part of many, and the way in which these misconceptions served as the basis of a broadside attack on science. Although the strictly scientific criticism of the content of the theories (on the part of scientists) was legitimate, stronger action (against a publisher) cannot be justified. All in all, a study of this episode is fascinating for the attitudes, prejudices, and ignorance that are revealed.

4.5   Disputed Scientific Advice in Public Issues

We turn now to rather more serious problems of scientific controversy, in matters of public policy. For each issue there is an extensive literature, and each provides an excellent subject for more detailed study than would be in place here. We shall therefore concentrate on the form of an issue and point to the types of questions that are raised.

Consider first the disputed association between smoking and lung cancer. By now the weight of medical evidence very strongly implicates smoking with an increased incidence of lung cancer although not all doctors yet agree with this. Establishing an association stops short of a full understanding of the details of the mechanisms involved. How can the remaining scientific questions be solved? Only by further studies, of animals and of men. Should effective legislative action be taken, to curtail the sale of tobacco products or at least further reduce their advertising? Should any such action be taken before the scientific evidence is conclusively established? What degree of scientific acceptance is sufficient to warrant legislative action? Taking no action implies the continuation of present sales practices. These questions are being posed here to show that practical questions hinge on scientific questions and may need to be decided even before the scientific certainty is 100 percent.

What about the use of chemicals for pesticides or for fertilizers or for herbicides? There is now increasing evidence that some of these chemicals produce adverse effects in laboratory animals. What degree of scientific certainty is needed before action should be taken to restrict sales? How about a new drug that is developed: At what stage should this be released for general prescription use, or for general sale to the public? And what of drugs already in use, whose side effects become well documented? Existing procedures permit extended litigation while a disputed drug continues to be sold. In all of these and in many other cases too, scientific doubts may have arisen regarding the safety of chemicals, drugs, or foods; the evidence may even be strong but short of being conclusive. Properly, the resolution of the scientific aspects of these questions will come as more research is conducted, and they can be properly classified as scientific controversies. One may consider that far more is involved, however, and that the often leisurely and unpredictable pace of normal science cannot be allowed to take its course, and this immediately places these questions in a class different from the controversies we have discussed earlier.

There is yet another area, related yet distinct. It is related in that major decisions need to be made even before all the scientific evidence becomes available, yet it is distinct in that the decisions will affect governments and international relations. We refer here to problems connected with nuclear weapons. By now much of the evidence has been opened to public discussion with the result that independent scientists can review much (but not all) of the details of proposed systems. What level of scientific and technical agreement is needed before a government decides that an ABM system will indeed work and that further policy can be based on it? How much scientific agreement is needed before a government decides it is safe to continue with a program of testing of nuclear devices, some below ground but some above? Here we find severe cleavages running through the scientific community. Given this and the resulting divergent opinions, how shall the government decide? What other factors enter?

Chap. 8 will deal with topics such as these, where science and politics mix. While we may patiently await the next issue of the *Physical Review* to read the outcome of some experiment, governments are sometimes not in such a fortunate position of being able to wait—or perhaps they should have to wait?

NOTES AND REFERENCES

1. Michael Polanyi, *Bulletin of the Atomic Scientists*, **13** (1957), 114. See also Warren O. Hagstrom, *The Scientific Community* (New York: Basic Books, Inc., 1965), Chap. VI.

2. James D. Watson, *The Double Helix* (New York: Signet Books, New American Library, 1969).

3. Thomas S. Kuhn, *The Structure of Scientific Revolutions* (Chicago: The University of Chicago Press, 1962).

4. See, for example, Boris V. Derjaguin, "Superdense Water," *Scientific American* (November 1970), 52; *Physics Today* (October 1970), 17; *Science*, **171** (1967), 167 and **172** (1971), 231.

5. L. Pearce Williams, ed., *Relativity Theory: Its Origins and Impact on Modern Thought* (New York: John Wiley and Sons, Inc., 1968).

6. Edward U. Condon, *Final Report of the Scientific Study of Unidentified Flying Objects* (New York: Bantam Books, 1969). Conducted at the University of Colorado under contract to the United States Air Force.

7. Immanual Velikovsky, *Worlds in Collision* (New York: Dell Publishing Co., 1967); see also Martin Gardner, *Fads and Fallacies in the Name of Science* (New York: Dover Publications, Inc., 1957).

8. *American Behavioral Scientist* (September 1963, June 1964, October 1964); *Bulletin of the Atomic Scientists* (April 1964), 38; *Yale Scientific Magazine* (April 1967).

organized science—
new roles and
decisions

chapter 5

As with almost all parts of western society today, science is organized. There are professional societies catering to all specialities, there are coordinating associations at many levels, and there are small groups whose restricted membership confers great prestige. In recent years, this interlocking framework of science has been the subject of study in its own right. The sociology of science, the political manipulations, recognition and awards, all of these aspects and many more have been treated at increasing length in books and articles. This is a fascinating area, filled with personalities, committees, structures, and (potentially and often actually) millions of dollars. These, however, are not the main concern here, but some understanding of at least the barest bones will be useful and will provide the basis for a better appreciation of the issues on which we shall concentrate.

For the most part, scientific societies have confined their efforts to immediate and obviously professional activities such as publishing journals and organizing conferences. Now there is pressure to have these bodies make statements and adopt positions on the behalf of organized science. Should such statements be made? If so, are there any restrictions as to the subjects that might be covered? These are some of the questions now being raised which will be explored in the present chapter.

For a start we can consider a professional society, such as the American Physical Society or the American Chemical Society. Such societies usually restrict their membership to those who are qualified either by training (university degree) or experience. Technical competence is a prerequisite, for along with membership goes the right to present papers at the regular meetings of the society. Societies usually publish journals, but membership does not remove the requirement for having papers submitted to review prior to publication. Conferences come in all shapes and sizes: mammoth national meetings with 20,000 attending or local sectional meetings with less than fifty. Some conferences are so large that they can be held in only a few cities that have sufficient hotel accommodation, while others are deliberately set in isolated spots to permit the maximum concentration on the part of the small group of specialists who have been invited.

At the international level, the International Council of Scientific Unions acts as an umbrella, coordinating the activities of other groups such as the International Union of Pure and Applied Chemistry, the International Scientific Radio Union, etc. Through these there can be organized the many and varied international conferences and the international collaborative efforts such as the International Polar Years or the International Geophysical Year.

Not formally incorporated but nevertheless having great influence in the progress of science are the "invisible colleges." This name was first used to describe the group that met in 1645 at Gresham College, Oxford to discuss scientific affairs. From this there was formed, in 1660, the society which became the Royal Society (given its Royal Charter in 1662). More recently the term "invisible college" has been used to describe the loose association of specialists who exchange preprints and personal communications and constitute a sort of brotherhood, usually international in membership and often exerting great influence. Through personal recommendation, it is from such groups that advisers to governments are sometimes drawn, and within such groups that candidates for vacant academic positions are sought.

As science has grown, so have more professional bodies been formed; more meetings have been held, more journals have been published, and new "invisible colleges" have sprung up. The guiding principle for the vast majority of these professional groups, both large and small, has been the furthering of science. Apart from the annual elections of office holders and an occasional change of bylaws, it has not been necessary to ask members to express their views, and the summation of members' views has not been assembled in order to be used as expressions of corporate opinion. (The well-known "nine out of ten doctors agree. . ." does not represent such an opinion.)

At one time science progressed quietly out of the public view, apart from spectacular matters such as a new 200-inch telescope or the discovery of some new wonder drug. While science did not attract much publicity, it nevertheless was steadily producing discoveries and applications that were more and more affecting the lives of millions. There is a tendency to overlook this part of science and to consider that the public impact dates from 1945 with the first detonation of a nuclear explosive. What changed then was the public awareness, and what has changed since then has been the rate of change inflicted by science.

We are accustomed by now to the recognition that science can be used to good or bad or debatable ends. Should scientific academies applaud the discovery of DDT or the invention of the laser? As a matter of fact they have. Moreover, medals and awards have come to those whose discoveries and inventions have clearly benefited mankind or have opened up new areas to science. How about the converse: Should science condemn certain other applications or discoveries? Even in our approach to successful or "good" science, we can see differences: We applaud the discoverer of penicillin, but the awards do not go to the pharmaceutical houses which have been able to mass-produce modern drugs. If we assume that there can be agreement on the definition of "bad" or "evil" discoveries or applications (or even without such agreement), what are the appropriate attitudes to adopt? The current attempts by some persons to brand scientists as wicked whose discoveries have been misused have not made an important distinction: If blame is to be laid, would it not be appropriate to aim at those who have taken the discoveries and made the applications rather than at the (as likely as not) innocent discoverer? Furthermore, in the expression of such opinions on scientific discoveries and their applications, what should be the role of the scientific academies and professional societies?

5.1 Collective Opinions

Professional science does not settle scientific disputes by an election. The numerical value for the speed of light is not settled by a referendum among the members of the American Physical Society or the Optical Society of America. Experimental evidence, as it accumulates, will alone settle the question. If scientists then are reluctant to express corporate views on strictly scientific matters, why should they do so on nonscientific or semiscientific issues where one might question their being any more expert than various other nonscientists? (One simply cannot imagine much value being placed on a statement such as: "Whereas 83.6 percent of the members of the Astronomical Society considered the moon to be made of cheese; and whereas only 14.7 percent was sure it was not made of cheese, while 1.7 percent expressed no opinion; now therefore be it resolved that the moon is probably made of cheese." This may seem absurd and yet, under the stress of political events, university faculties and professional groups have indeed adopted resolutions of this form even though outside the areas of their professional competence, without making any attempt at critical study.)

What value would there be to an institutional opinion, representative of the majority of those members of some professional body who happened to participate in a referendum or election? Perhaps there may be some occasion when such an opinion will carry great public weight. For instance, if the overwhelming majority of the membership of the American Medical Association were to condemn smoking as an important contributing factor in the incidence of lung cancer and other diseases, then the effect upon the public would be impressive. Such an expression is not simply the result of a vote and a tabulation of opinions; it carries with it the weight of knowledgeable authority and the implication that these experts, after weighing the information available, have come to the conclusion which they then state. It is difficult to see how a scientific body could exercise its expertise, its cool unbiased consideration of data and issues were it to express itself on some matter not strictly scientific. Not all of the public is critical in examining corporate opinion, however, and corporate opinion does tend to carry public weight, more weight, indeed, than if the same people simply published a statement saying "the following 2973 members of the American Society of Blank think as follows. . . ."

Can scientific responsibility be reconciled with a refusal to permit scientific societies to express the majority views? (The broad issue of social responsibility will be examined in Chap. 10. Here we would like to confine our attention to the role of the professional societies.) What would be lost if societies did express their opinions? A great deal, I suggest. In the extreme, we would find the presidents, secretaries, and committees of scientific bodies being elected on the basis of platforms and planks (probably including foreign policy), the societies would be split in ways that would reflect the outside political allegiances of the members, and science would no longer rise above other partisan issues. To avoid voting on certain issues in certain places is not to ignore the issues, but rather to deflect their advocacy to some more appropriate group or arena. Membership in a professional scientific society does not (nor should it) require conformity in political or social views, and there seems no reason for the majority of such a group to attempt to speak for all, thus imputing those same views to a dissenting minority. Other ways have been found to provide for expression of strongly felt views and to carry them further in the form of lobbying, and the Federation of American Scientists is just such a group, giving voice to its views on many issues. Very recently, the American Physicists Association has been formed, distinct from the American Physical Society, and having a range of interests beyond the narrowly scientific.

There is at least one more reason why the expression of a majority view, on matters more political than scientific, may be undesirable. Major societies, such as the American Physical Society, have an international membership, and by now this extension of cultural contact has wide support. The expression of a majority view may indeed be representative of the local membership, but could well be a considerable embarrassment to the foreign members. Perhaps there may be some matters of such international and critical importance that national borders and differences are transcended, matters where an overwhelming voice of conscience is needed, although experience suggests that there will usually be disagreement over just which topics fall within this category.

With this preamble it is instructive to examine a case history in which just such an issue was brought to the American Physical Society in the form of an amendment proposed for the Society's Articles.[1]

Resolutions.—The members may express their opinion or intent on any matter of concern to the Society by voting on one or several resolutions formally presented for their consideration . . .

In terms of the procedures proposed, one percent or more of the membership could initiate such a vote by placing a topic before the Society. The proposal was circulated to the members of the Society towards the end of 1967, and received editorial comment in the December 1967 *Physics Today*. (*Physics Today* is an official publication of the American Institute of Physics, and carries news reports, book reviews and topical articles but is not a journal for original scientific publications.) This editorial is reproduced in the appendix together with a selection of the letters that followed. From the correspondence it was clear that there were strong views on both sides, based on many different reasons. In the mail ballot of the Society's members, 55 percent voted, and the proposed amendment was defeated by 9214 votes to 3553.

Although the ballot decided the matter within the Society, there were clearly many who felt that some organization was necessary through which views could be expressed, and when many members attended the February 1969 meeting in New York, an attempt was made to form such a body. The notice calling for this meeting is also reproduced in the appendix. No immediate action seems to have occurred, but about a year later an independent body, the American Physicists Association, was formed. This new group emerged partly in response to the earlier discussions but also to fill a need that the younger physicists perceived as the employment of physicists was becoming much more difficult. Some felt that the older bodies had neglected the employment aspects. At the time of this writing, it is too early to say what effect and direction this group will have. It is, however, interesting to observe that there seems to have been no similar movement yet within the professional ranks of any of the other areas of science or of engineering.

Although there is great reluctance to see a professional society make pronouncements on nonscientific matters (as shown by the APS vote), there has been recognition of the widely felt need for the provision of a place for discussion where issues may be aired without any vote being taken. Both the Physical Society and the American Association for the Advancement of Science (AAAS) have included panel discussions of topical matters on their programs, and the AAAS has been doing so for several years. There is surely room for this idea to be taken over by other scientific societies, and one might even look forward to other professional bodies participating. For instance, an occasional invited speaker or discussant at the Modern Languages Association from the sciences could be part of an exchange of mutual interest. But this is drifting away from the main theme, which is that the role of the professional association is being broadened and that it is reasonable to undertake some experimenting.

Outside the usual ranks of organized science, groups have been formed for specific and usually short-term ends. For instance, during the 1964 presidential election campaign, an ad hoc group, "Scientists and Engineers for Johnson and Humphrey" campaigned vigorously. It is not clear that this political activity has had any permanent effects on science, although this might be the case if such groupings were to emerge every four years with the same intensity that marked the 1964 campaign.

If we prohibit scientific societies from making political statements and insist that such expressions come from other groups, have we ensured that science will be politically neutral? Does this completely exclude all lobbying by organized science. Alternatively, might we expect lobbying on matters directly affecting science? What should be the position of the societies or the National Academy of Science when the proposed budget for the National Science Foundation is being reviewed in Congress? Is this an area where the Institute of Physics should request to testify at appropriate hearings, or should scientific testimony be left to chance with individuals appearing on their own behalf carrying only the prestige of their reputations and positions? Alternatively why should organized science refrain from the expression of its views? After all, in the hearings and debates on the science funding appropriations, there will often be many views that are not firmly based on an understanding of what science and technology can and cannot do, presented by persons outside of science. Who then speaks for science? Even in the absence of expressions of scientific opinion and values, decisions *will* be made—if not earlier, then at least by Congress when it gives or denies authorization. The Administration and Congress often seek advice; should science wait to be asked? Some of these questions will be discussed in Chap. 8; nevertheless, they are presented here because they are asked of organized science.

5.2 Scientific Conferences and Politics

There is a very different problem that has confronted organized science in recent years. Where should a particular conference be held? If it is held in a country with a totalitarian regime, does this imply support for that government? In 1969 a group of molecular biologists raised this question in a letter to *Nature*. The fourth NATO Advanced Study of Molecular Biology, internationally attended, had been held in Greece. Many scientists disapproved strongly of the Greek military government. Should they have attended as a token of their solidarity with Greek scientists or should they have simply attended, not trying to use their presence in any way other than in the pursuit of science, or should they have declined to attend? And declining, should they have tried to get the meeting cancelled or moved? With a wide range of possible courses of action, who should make the decisions? Only those scientists invited to attend? NATO? or all members of the affiliated professional bodies concerned with the meeting?

There were editorial comments in *Nature*, and these together with some letters are reprinted in full in the appendix. The issues are well set out, and it would seem that, with certain precautions and guarantees, it should be possible to hold meetings almost anywhere. There has been a strong trend in recent years to present political issues as though they permit only two solutions rather than a solution chosen from a continuum. Here too there is a tendency to oversimplify through the presentation of a false dichotomy: That to hold the meeting in country X implies support or condoning of government X, whereas shifting or holding the meeting elsewhere implies condemnation. It should be noted that foreign scientists and other intellectuals have at times experienced severe difficulty in obtaining visas to visit the United States; on the other side, there have been meetings in Russia from which Israelis have been excluded, yet there have been conferences in Hungary to which attendance was completely open to the extent of admitting scientists from countries with which Hungary had no diplomatic relations. In the planning of a major meeting, therefore, there are many factors to be weighed, and it seems unlikely that there is any single rule that will serve as a universal and satisfactory guide. In addition, there are still individuals who feel quite strongly about not attending meetings in certain countries, but this is a matter of personal decisions and should not impose restrictions on the remainder of science.

A similar question arose after the tumultuous demonstrations that took place in Chicago in 1968 during the convention of the Democratic Party. Many professional societies (not only in the natural sciences) regularly schedule major meetings in Chicago, and there was considerable sentiment in many groups to boycott Chicago as a meeting place. The American Physical Society together with the American Association of Physics Teachers had already set their joint January meeting for Chicago in 1970. The membership was polled after it had been proposed that the meeting be held elsewhere (see the appendix for statements that were circulated), and the majority decision supported the Council in going ahead with the meeting as planned.

Other societies (philosophy and history) decided to shift meetings that had also been scheduled for Chicago. Does this mean that physicists are less politically aware, or more conservative, or more businesslike? There is room for endless speculation on the combined meaning of the two referenda to physicists, but the issues remain in certain ways and will doubtless rise again as events provoke strong feelings.

By now, it should be clear that organized science will be faced by the need to make decisions that can (and often will) be interpreted as politically inspired, all denials notwithstanding. It is doubtful whether it is either possible or desirable to try to formulate an explicit and exhaustive set of rules by which all future questions can be easily settled. Changing times and social attitudes will surely produce changes in the way in which scientists approach and view their professional bonds, and accordingly in the balance they strike between these bonds and those which they feel to the larger society.

NOTES AND REFERENCES

1. *Bulletin of the American Physical Society,* **12** (1967), 1110.

the university:
a home for research

chapter 6

In preceding chapters, the discussions have covered several aspects of the way in which science is organized and conducted; in the present chapter, attention will be given to the important question of where it is carried out. A very few large industrial corporations have research laboratories of renown where pure research has been encouraged for many years: In 1937 Clinton Davisson was awarded the Nobel Prize in physics for work carried out in the Bell Telephone Laboratories, and in 1956 John Bardeen, William Shockley, and Walter Brattain (also from Bell Laboratories) shared the Nobel Prize in physics for their development of the transistor. At the General Electric Research Laboratories, Irving Langmuir's research brought him the 1932 Nobel Prize in chemistry. These are exceptions, however; industrial research has been very largely concentrated in applied science. While government laboratories, such as the National Bureau of Standards, the National Center for Atmospheric Research, and the Oak Ridge National Laboratory do undertake some basic research, their main work, too, is applied.

Most pure (or basic) science in the United States today has its home in the universities, and this concurrence will be the focus of the present chapter. Increasingly the scale of modern scientific research has led to strains in the universities, and there has been continued argument on the appropriateness of research in the university. In considering the housing of science, we therefore need to examine its relationships with the university, and we shall need to digress slightly at points along the way in order to describe some of the educational aspects insofar as they affect or are affected by the presence of research on the campus. But the center of our interest will still be scientific research; there is already a vast literature dealing with the structure of universities and with suggested university reform, which is an interesting subject in itself.

Universities are unusual institutions, closely comparable to no others in our modern society and not fully understood either by their inhabitants or by the public. They are often regarded as a sort of mixture of educational cafeteria and service station, or as agencies for the solution of many of society's problems. At other times the university is viewed as the source of unrest and revolution, the fount of all social ills. In the midst of this, it is not surprising that science on the campus should attract attention.

Academic science today finds itself under attack from two sides. Some critics complain that many faculty members are so engrossed with their research and associated activities (consulting, advising) that their participation in undergraduate education suffers. Other critics complain that too much university research consists of abstruse (and sometimes excessively expensive) topics that are really luxuries, and that more effort and money should be diverted into working on problems which are of immediate concern to society—problems such as those dealing with our cities or health or the environment. Should the universities abandon all research to industrial and government laboratories, and rather concentrate their efforts and talents on education? This is a division of skills which is indeed followed in some countries, where the universities do the teaching, and the basic research is virtually confined to government institutes. Alternatively, is there good reason to retain research as a constituent of the university? If we do retain it, should there be changes, either in the scale or in the approaches?[1]

There are over 2000 institutions of higher learning (i.e., post high school) in the United States today. If we define "research" in the university context as those studies that can ultimately lead to a Ph.D. degree, then we find ourselves considering a far smaller number of places: While the Ph.D. is awarded by approximately 200 institutions today, the larger 100 of these award nearly 95 percent of all Ph.D.'s.[2] It might seem then that the problem is relatively small, but this is not in fact the case. Those 100 universities are the largest and most prestigious, for they include the Ivy League schools, the other major private universities (such as Chicago and Stanford), and the great state universities. It is these universities that set the standards against which all are judged and to which many aspire, and their influence extends far beyond their numbers. The very large number of institutions which remain include the many small liberal arts colleges, each having an average enrollment of less than a thousand students, the growing number of state colleges which do not yet offer the Ph.D. or else have only very new Ph.D. programs, and the rapidly increasing number of junior or community colleges.

Recognizing exceptions, we can itemize the major differences between typical four-year colleges and the Ph.D.-granting universities. Usually the universities have a larger percentage of faculty members who have Ph.D.'s themselves. Universities are also more selective in the students they admit. These two factors usually lead to more demanding curricula from which a higher proportion of graduates will continue towards a Ph.D.

In theory, the teacher at an undergraduate college who is unencumbered by research will be able to devote himself completely to undergraduate education. He can spend more time teaching. With higher teaching loads, it should be possible to have smaller classes and more individual attention on the part of faculty instead of leaving a major part of many freshman classes to instruction by first-year graduate students. More time can also be made available for advising and counseling, and in smaller schools it is possible for a greater proportion of the student body and faculty to know one another.

In contrast, a faculty member at a university (using this term to denote those institutions where active research is expected of most faculty) finds his time fragmented. In addition to taking his turn with undergraduate teaching, he is expected to teach graduate seminars and supervise graduate students, and his own research progress will probably be a deciding factor when he is considered for promotion. The university is probably also large enough so that he will find himself serving on committees, which proliferate these days. If he is well-known in his professional field, he may be in demand for governmental advisory panels and/or industrial consulting. What time is then available for attention to undergraduate teaching? Against this, one might claim that he can bring a deeper insight to his undergraduate teaching and can draw on his current research for interesting examples with which to illustrate even a routine course.

As is the case with a good political cartoon, there is enough of the truth in both of these caricatures for us to recognize the characters, but the proportions have been distorted, and what has been omitted may be just as important as what has been included. It is necessary to probe deeper to obtain a more balanced view of the role of research in the university. Undergraduate education is a good place for some starting comments.

6.1 Undergraduate Education

There is something very artificial about science textbooks. They represent the codification of successes in science, and few of the wrong turns are included. Pedagogic clarity would certainly not be aided if one were to attempt to include a detailed chronological account that followed every twist and turn, some of which have led to our present knowledge, some of which were of great heuristic value, and others of which proved to be unproductive diversions. Even though the exploration of these aspects of scientific progress can be fascinating, and even though it is precisely these detailed twists and turns that constitute much of the activity of normal everyday science, they are better appreciated after having achieved a firm foundation in the present knowledge.

Both for reasons of clarity and economy of time, the presentation in science textbooks is, therefore, very different from the way in which that science was originally reported, in journals or in treatises. With experience clearer ways can be found to present older ideas; style and emphasis change with time, and the textbooks reflect these changes. The overall result, though, is that too often little or no impression is transmitted through the texts to show that science is still alive, progressing, and changing. Instead, many texts present the artificial picture of science as a still life, the lighting consciously arranged and the highlights and shadows carefully contrived. Happily there have been major attempts to rectify this picture in recent years, and revised curricula and texts have emerged in the sciences and mathematics through the combined efforts of high school, college, and university people.

Major contributions to many of these revised curricula have come from prominent scientists who have also continued with their own research work. Stimulation in the reverse direction should not be overlooked; a research scientist who is engaged in undergraduate teaching should be continually examining his own understanding. Although it is certainly good to have periods for research work uninterrupted by the problems of undergraduate homework and examinations, most scientists feel the need for continual stimulus such as comes from the unexpected question in the middle of a presentation which the lecturer would have expected to be easily grasped.

The presence of active research brings with it some other major advantages to the university. A research faculty needs a good library with both a range of books and the most important current periodicals. Especially at the more advanced undergraduate levels, students gain by being able to go back to some original sources and by having access to the current literature. This is of particular importance when coupled with the trend to involve undergraduates in research activities in which they acquire some skills and, much more importantly, can find a real interest by working alongside those with much more experience. With the flexible arrangements at many universities, advanced undergraduates can also be encouraged to take selected graduate courses.

There can be no denying the fact that some university faculty members have made little or no effort in undergraduate teaching, but we should also recognize the other side of the coin, namely that there are indeed many who are active in this way. Similarly in colleges there are many dedicated good teachers but there are also those whose knowledge is out of date and whose resources are small. While there is good teaching and bad teaching at both colleges and universities, neither has the monopoly of either category. On balance, there would seem to be greater possibilities at the universities for superior teaching, which is not to say that these have always been recognized or exploited.

6.2   Graduate Education

So far we have been concerned with the undergraduate years, and here all colleges and universities are included. When we turn to graduate education, our attention is narrowed. Where should graduate education be pursued? Should it be confined to purely graduate institutions with the possibility of an alternative such as an apprenticeship at a national laboratory, devised to take the place of university research education? Should there be a clear and rigid separation of graduate and undergraduate education? To some critics, research is such a distraction from undergraduate teaching that they have seriously proposed this intellectual apartheid. It is probably fair to say that few faculty members would support such a separation of teaching and research as most prefer to meet both undergraduate and graduate students. At present, research participation forms an essential part of the Ph.D. program in addition to the formal graduate courses. With basic research so heavily concentrated in the universities, there is effectively no other place where a graduate student can do his research, apart from a few who may through special arrangements carry out their work at some government laboratory. Unless there were to be a truly massive rearrangement in the basic research patterns in this country, it seems quite unrealistic to consider separation of this research from the universities. If anything, the tendency is in the other direction; the Rockefeller Institute, which has had a distinguished record of research in the biological sciences, has become Rockefeller University even though it confines itself to graduate programs and awards only Ph.D.'s.

There is another reason for housing research in the universities: Freedom to choose the subjects of research and to publish the results. This freedom can be severely curtailed in government or industrial laboratories; in universities academic freedom is more than simply an empty phrase, and the way is open on the campus for individual initiative even if it results in the occasional choice of some useless or foolish subject. Freedom to choose to undertake research in an unpopular area may appear to be more important to those in the social sciences, where topics of current interest may offend puritanical or politically conservative views, but this freedom is still important in the natural sciences, especially with the pressures now increasing to undertake "useful" work.

A problem that may confront the university scientist is the matter of his allegiance. A working scientist has strong professional ties to his colleagues in other universities. Faculty members are mobile (or were until the very recent and serious reduction in federal research support); the charge has frequently been made too that many scientists feel no strong tie to their universities. It is a resting place for the moment, but an attractive offer from some other institution can shake them loose. Perhaps all that can be said is that this is probably true for some, but not for most, and that this mobility is by no means confined to the sciences.

6.3   Some Problems Related to the Scale of Research

So far, we have been concerned with the effects of research on the graduate and undergraduate education, and with the freedom of choice of faculty scientists. These are the more obvious effects of the presence of research, but there are others, too, not as obvious, that pose severe problems for the university and that need to be at least mentioned.

One set of problems that may arise is connected with the scale of modern research. In almost all cases, science is now too expensive to be supported out of university funds. National agencies such as the National Science Foundation, the National Institutes of Health, the offices of scientific research within the Department of Defense, and the National Aeronautics and Space Administration provide almost all of the support with a diminishing proportion coming from private foundations.

There can be no brief survey of university scientific research with any pretensions to completeness. It is a vast enterprise with the spectrum extending from the isolated professor who works away for many years on a single problem to a team of ten to twenty faculty and graduate students who collaborate with half a dozen similar groups at other laboratories. There has been widespread discussion of big science and little science and the problems of the financing of science on a national scale. Our main interest in this chapter is the relation of the research to the universities, and the books by de Solla Price, Weinberg, Orlans, and Brooks (cited in the bibliography) can be recommended to those who wish to explore some of the other aspects.

Individual scientists are free to invent interesting projects and then to try to get support from some agency. Broad policy decisions on the part of the federal administration, the Bureau of the Budget, and others can seriously affect the support of whole fields of research and thus the ability of a scientist to get support for his proposed research. Scientists must learn to recognize the presence of strong tides in the flood of public money.

Some scientists do more than merely observe these tides. They can divine the presence of underground reservoirs from which streams will soon come bubbling up to the surface. As with all trades, science has its entrepreneurs. An enterprising scientist can arrive back on the campus from a visit to Washington with the promise of a large amount of money for the setting up of some new interdisciplinary institute. How should the university respond to the prospect of such sudden munificence? Inevitably other university resources will have to be diverted to help support this new entity since external funds do not provide support for all the true costs or do not do so for more than a brief initial period. The diversion of internal university funds implies a shift in the internal priorities from those which the university would otherwise have chosen. So, how should the university respond to the prospect of a new institute? Should the decision be entirely administrative or should there be faculty involvement? (and, moreover, should there be student participation?) Does this sort of review constitute an infringement on the freedom of the scientist to undertake whatever research he wishes for which he can find financial support? There is no standard prescription for dealing with this problem, which faces all large universities. To pursue it further would take us into the jungles of university governance, now the subject of strident controversy at many campuses where nonnegotiable demands conflict with the defense of traditional structures. The main point we need to see here is that largescale science brings problems to the campus, and the complete freedom of a scientist to pursue his interests may be subject to limitation when the scale of his proposed project becomes very large.

6.4   "Secret" Research and Military Sponsorship

Our problems are not yet over. Since World War II, many universities have sponsored research for the military which is classified; i.e., the work is secret, and those engaged in it must have gone through a security clearance investigation. Sometimes this research has been housed on campus; in other instances separate buildings have been used, separate institutes or laboratories developed, and links with the university stretched. Some universities chose never to house secret research. With the growing unpopularity of the military (a direct outgrowth of the war in southeast Asia) shown by many students and faculty, all links with the Department of Defense have come under scrutiny, and secret research projects have been most heavily attacked.[3]

At those universities that have firmly set themselves against secret research on the campus, the view has prevailed that secret work in a university is inherently antithetical to the very nature of the university—free and open learning and enquiry. How can a graduate student be engaged on secret research and not be able to publish his results and thesis? How can one be sure that secret research receives the same searching scrutiny that open publication ensures for other work? These have been for many sufficiently compelling reasons for avoiding secret research. What a faculty member does in his own time (such as during the summer) is his own business, however, and this need neither compromise the open nature of the university nor his ability to assist with secret projects.

Others have argued that the defense of one's country should not be considered with such contempt, and that the members of a university have an obligation to do their share.[4] One answer that has been given has been that the government should provide facilities for the work it needs, engage the scientists and consultants it feels necessary, but not interfere with the independence and integrity of the university. There are various ways in which government research can be pursued, but the universities are the only institutions in society engaged in teaching and research, and if this role is compromised there is no other institution to take over.[5]

Apart from secret research, should the university and its scientists accept *any* funds from military agencies, if the funds can be used for open research, chosen by the scientist with no restrictions on the publication of results? Is there some reason why industrially related research can be carried out in schools of engineering and business administration, while the other half of the military-industrial complex is denied this access to the campus? Perhaps a reasonable distinction might be possible on the basis of secrecy. A strong case can be made that secret work does not belong on campus (at least in time of peace) and that such work, if needed, should be performed elsewhere. It is, I feel, much harder to argue that open research, which could just as well be supported by the National Science Foundation (NSF) should be rejected simply because a military agency chooses to sponsor it. In fact, there will often be joint support, where neither agency alone is able to provide the full costs.

At one time very large amounts of research money came from the Office of Naval Research (ONR), the Air Force Office of Scientific Research (AFOSR), and the United States Army Research Office (ARO). Without their foresight and intelligent support, science in the United States would have been very different in the years before the NSF was established or funded at its present level. While science has more recently also benefited via NASA, the remarkable growth of science in the United States owes much to these military agencies. Basic science was generously supported. Some areas, such as nuclear physics, might seem to have been chosen for their connection with weapons systems, but the scientific results were still openly published. There was another reason for supporting basic research; The production of trained scientists who could be called upon in times of national emergency. This was considered sufficiently important so that the immediate military relevance of their research was not required. More recently, this attitude has changed under pressure from Congress, and military support is becoming more "mission-oriented."[6]

What then should be the position of the university and its scientists toward military support for their research? This question has been the center of bitter controversy at many universities. Nothing can force a scientist to undertake work sponsored by the military, and many have felt that the decision should be left to each scientist provided that the work is not secret. Others have expressed the view that by accepting any support from the military, a scientist is at the very least condoning the actions and policies of the Department of Defense. Here again we encounter political differences in the approaches to this issue, and these will be resolved, if at all, in different ways at different universities with no sign of any national policy at present.

The secret projects that we have been describing are those whose general existence (but not the technical details) may be known. In the past few years, two other kinds of secrecy have come to light. For instance, at one university, there was highly secret work on chemical and biological warfare. When this became known, the outraged and widespread university sentiment forced the cancellation of the project. In another case infiltration by the Central Intelligence Agency (CIA) was discovered which, in turn, cast doubt on the good faith of many other projects where it had been assumed that social scientists were engaged in field studies for purely scholarly purposes.[7] A different situation existed at the Massachusetts Institute of Technology (MIT) where its subsidiary Instrumentation Laboratory developed, among other things, highly sophisticated guidance systems, some of which were used for manned space-flight vehicles, others for Polaris-guided missiles. Strong protest has forced the separation of the Instrumentation Laboratory from MIT.

It is clear that the military feels that it needs the scientific help it receives from universities, and has consequently been unhappy about attempts made by Congress to restrict military research support to matters of obvious and immediate interest. To the military appropriation bill for 1969–70, there was added the so-called Mansfield amendment, which imposed severe requirements of immediate usefulness. This was strongly opposed by both scientists and those in military research offices so that similar but much weaker restrictions appeared in the 1970–71 military budgets for research.[8]

Most of our discussion has centered on problems of secrecy in military research on the campus, but we cannot leave this subject without looking into another matter of great concern to universities. Is there an inhibiting effect on the political activity of those whose salaries or research are dependent on the military? Generally, no. Although some scientists have preferred to seek funds entirely from nonmilitary agencies, others have accepted military support and still remained outspoken in their expression of political opinion. An interesting incident involving some mathematicians is so unusual as to warrant mention.

An advertisement in one of the professional mathematics journals read as follows:[9]

> Mathematicians: Job opportunities in war work are announced in the notices of the A.M.S., in the Employment Register, and elsewhere. We urge you to regard yourselves as responsible for the uses to which your talents are put. We believe this responsibility forbids putting mathematics in the service of this cruel war.

This notice was published several times and attracted over 200 cosponsoring signatures. A clumsy attempt was made within the Army Research Office to press some of the signatories into terminating their military research support, and it was suggested that those who did willingly accept military funds might find this hard to reconcile with their support of such an advertisement. More temperate views seem to have prevailed, however, and this episode passed with no further actions having been taken.[10]

In many ways it may seem that the contents of this chapter have strayed far from the central theme of the conduct of science. However, such a large fraction of basic research is conducted today by university scientists that it has been necessary to digress into the topics covered here in order to have a fuller understanding of the main issue. The scale of science can bring problems, as can military sponsorship, and care is essential if the university's values and freedom are to be preserved, but overall it would seem that the universities have much to gain by housing scientific research.

NOTES AND REFERENCES

1. For discussions of research and the universities, see, for example: F. A. Long, *Science*, **163** (1969), 1037; H. Orlans, *Science*, **155** (1967), 665; V. F. Weisskopf, *Bull. Atomic Sci*, **21** (1965), 4; W. K. H. Panofsky in H. Orlans (ed.), *Science Policy and the University* (Washington, D.C.: The Brookings Institution, 1968), 189.

2. For comprehensive surveys of graduate education, see: Allan M. Cartter, *An Assessment of Quality in Graduate Education* (1966); and Kenneth D. Roose and Charles J. Anderson, *A Rating of Graduate Programs* (1970); both reports published by the American Council on Education, Washington, D.C.

3. A chronicle of the discussions on this subject can be easily seen by turning through the pages of *Science*, especially during the period 1965–70.

4. E. Chain, *New Scientist* (Oct. 22, 1970), 166.

5. For a discussion of some broader aspects to secrecy and research, see *Science*, **163** (1969), 787, where there is published a report of the Committee on Science in the Promotion of Human Welfare, of the AAAS.

6. See, for example: *Science*, **158** (1967), 1032.

7. *Science*, **156** (1967), 1583.

8. *Science*, **166** (1969), 1386; **167** (1970), 1473; **169** (1970), 1059; **170** (1970), 613.

9. *Notices of the American Mathematical Society*, **15**, No. 1 (January 1968), and **15**, No. 5 (August 1968).

10. *Science*, **161** (1968), 1225.

the application
of science:
relevance

chapter 7

7.1 Some Historical Aspects

The cry for "relevance" is heard on many sides today, and science is under great pressure to help in the renovation of our cities, to cleanse the environment, and to provide new wonder drugs or cures. Basic or pure science is then viewed with disapproval as though it were not contributing to these worthy or essential social needs. In fact, the pattern of governmental support of science is shifting so that applied science is now receiving far greater support, not solely on its own merits but partly at the expense of basic science. Generally, the relation of science to technology is complex and is not widely understood. A clarification of this relation will be the purpose of this chapter.

Science is far more than an exotic hobby of a fortunate few. Despite the fact that it appears isolated at times, science has a habit of providing utilizable information. Whether these uses are for good or bad is quite another question, long recognized, but the findings of science do get sifted and find many applications although the delays between discovery and use vary greatly. While some discoveries have been promptly exploited, others have yet to find a use. With the aid of science, new machines and processes have been developed which not only help to improve the quality of living but also enable science itself to open up new fields.

In earlier times inventions were generally made without a full understanding of the basic scientific principles. Clearly the bronze and iron ages developed without a quantitative knowledge of the structure or properties of metals. Empirically rules might have been developed for the handling of materials or the improvement of inventions, but the underlying scientific facts generally were not known. Today this is far less common. As the historian Lynn White has so well put it: "Science had been a largely speculative effort to understand nature, whereas technology was an exclusively practical attempt to use nature for human purposes." Science and technology today have a symbiotic relationship, and the division between them is not nearly as well defined as the popular view would have it.[1]

Basic science has now progressed so far that there can be little applied science that is conducted on the hit-or-miss basis of earlier days when experience and intuition were better guides. We can see another important change: Public expectations of science and technology have changed markedly. In an earlier age of science, isolated discoveries were adapted for more general use but without thought of initiating other changes that would reach far beyond the fields of immediate application. (It is unlikely that Michael Faraday in his researches on electricity could foresee or be stimulated by the prospect of having electric toothbrushes.) We see a clear change at the time of the Industrial Revolution when the prospects of unlimited mechanical power replacing human and animal efforts led to almost utopian visions—of greatly reduced human labors, improved living conditions, and plentiful goods. Today we seem to have travelled almost exactly halfway around the circle from that time: Science and technology stand accused, as society surveys the accumulated debris of its successes—old automobiles, discarded bottles and cans, and pollution in air and water.

Before we discuss some of the present problems raised by the demand for relevance, it would be useful to remind ourselves of a few historic examples. Many of the great scientists over the years were men of great practicality, dividing their time between discovery and gadget invention. These men displayed an interest in what went on around them and devised practical solutions for the problems to which their attention had been directed. Many took an interest in machines of war, and in fact military consulting has a long and honorable history. It is only in the past few years that scientific association with the military has become unpopular in this country. It is well to remember that in different circumstances military applications of science were viewed quite differently (as witness the involvement of American science in the Manhattan Project to produce nuclear weapons in World War II).

Leonardo da Vinci is famous for his sketches of war machines, some of which foreshadowed the modern tank. He recognized that there could be circumstances that might make it necessary to become involved in military work:[2]

> *"When besieged by ambitious tyrants, I find a means of offence and defence in order to preserve the chief gift of nature, which is liberty."*

He also saw that inventions could be used in ways neither originally conceived nor to the liking of their originator, and his comments, while perceptive, are at variance with ideas generally held today, when we prefer to
work at the control rather than the suppression of devices that lend themselves to ill use. Da Vinci commented thus on his ideas for a submarine:[3]

> "How by an appliance many are able to remain for some time under water. How and
> why I do not describe my method of remaining under water for a long time as I can
> remain without food; and this I do not publish or divulge, on account of the evil
> nature of men, who would practice assassinations at the bottom of the seas by
> breaking the ships in their lowest parts and sinking them together with the crews
> who are in them."

Galileo is known for his careful experimenting, probably the first modern
scientist in this sense. Besides his contributions to science, he also improved
upon the design and construction of a variety of instruments. While resident
in Padua, he frequently visited the Venetian arsenal, and the opening sentence
of his great dialogue "Two New Sciences" refers to this:

> "The constant activity which you Venetians display in your famous arsenal suggests
> to the studious mind a large field for investigation, especially that part of the work
> that involves mechanics, for in this department all types of instruments and machines
> are constantly being constructed...."[4]

Not only was Galileo interested in these devices, but he displayed an
initiative that is still to be seen today, as in the famous Route 128 around
Boston where small companies have sprung up, drawing their inspiration
and often their founders from the technically expert at local universities.
Galileo went into business, employed a technician, and made and sold a
"geometrical and military compass," a device that he had considerably
improved. He also published a booklet for its use and pursued through the
courts an individual who plagiarized it.

As we follow this line through a variety of examples, we find a general
similarity in the circumstances: A man of ingenuity and invention, advancing
science and also devising new instruments or improving old ones. Some held
university positions; some were employed by the government (or had as
patron a duke or local authority); some were independently wealthy. What
is common to many of the applications and inventions of this time is their
isolation. They represented individual improvements, which occasionally
even lead towards devices of great utility, but it was rare that their effects
became widespread (especially on a short term), nor were there expectations
or pretensions of great social effect on the part of their inventors.

With the upheaval we now term the Industrial Revolution, the situation
changed. We find among inventors a conscious recognition of their role
in producing social change. The British Association for the Advancement
of Science, formed in 1831 and the Great Exhibition of 1851 are only two
examples of the manifestations of the hopes held by the public for technology

and its results, a technology which, in turn, was fed by science. Public interest was encouraged, was wide, and went along with the Victorian belief in "progress," improvement of the lot of man, and the missionary zeal that found expression in the colonies.

This wide interest and these hopes stemmed from the uses to which mechanical energy was being applied, but the quantitative definition of the term "energy," in the form we use it now dates only from 1847. Today we are all familiar with "energy" even though most of us may not be able to define it precisely. We have some understanding that it is related to the ability to perform arduous tasks, and we recognize the need for fuels in the production of energy: Food for humans and animals, gasoline and coal for engines. Our modern technological society is based upon the use of energy to such an extent that we must now deal with the undesirable side effects of energy production.

"Energy" also plays an important role in the formulation of precise scientific theory. The state of an atomic or molecular system is described in terms of its energy or the change in this energy when radiation is absorbed or emitted; the absorption of solar energy is involved in the growth of plants. Before the idea of energy had crystallized, the central concept in physics was that of force, and this we owe to Newton whose laws of motion describe the ways in which an object responds to the forces applied to it. Modern quantitative science developed from that formulation of the laws of motion; dynamics, which deals with masses, velocities, and forces, was developed to a high level of sophistication. Gradually the idea of mechanical energy was developed. Independently other branches of science progressed, and among these was the study of heat. The various branches of science along their parallel but separate paths defined their own units: Pounds, feet, and seconds for dynamics, temperature scales in degrees for heat. A unit of thermal energy was also defined in terms of measurable changes in temperature. One of the major advances in physics came when James Joule, as a result of great experimental care and precision, showed that there is a fixed relation between the dynamical and thermal units of energy. Although originally defined separately, these units do represent a definite amount of energy. If one unit of mechanical (dynamical) energy goes into producing heat, then a certain amount of heat energy is always produced and no more, no less. For example, if one rubs two surfaces together with a certain force, one is using a certain amount of mechanical energy. If all other conditions are the same, then one will always produce the same heating effect in the rubbed objects. We use this in striking a match, and the heat generated by the rapid rubbing of the match head on a rough surface is enough to set off a chemical reaction in the match head which leads to combustion; we use this when applying automobile brakes, converting the vehicle's forward mechanical energy to heat energy (in the brakes, drums, or discs).

Although some correspondence between mechanical and heat energy had been suggested earlier, Joule's work (published in 1843) was the first to show an unvarying and exact proportionality, and it established the first clear link between what until then had been studied as separate parts of science. This in itself is important, but we find far more emerging when we inquire into the background to Joule's experimenting.

Joule was the son of a wealthy brewery owner and lived in Manchester in the English industrial midlands. Perhaps it was chance that caused him to experiment as he did, but it would seem more reasonable to think that he must have been influenced by the rapid industrialization going on around him. In fact, when we take a broad view of the scientific and technical progress around that time, we shall see their close connection, strongly influenced by social needs.

Consider, for a moment, the situation by the middle of eighteenth century. Wood was an important material both for fuel and for construction (for example, for ships and houses). Increasing demands for wood were leading to severe shortages, but alternatives were few. Coal might be used for fuel, but its mining was not easy. The easily stripped surface deposits were rapidly exhausted, and deeper mines soon ran into problems. Water had to be pumped from considerable depths, and the coal had to be raised to the surface. Even then the coal could not be used in simple ways for fires by which iron was converted to steel, because impurities in the coal led to contamination and highly undesirable properties in the resulting steel. Wood was thus preferred as a fuel, but it was in short supply. A major part of the solution to this circle of dilemmas lay in the invention and development of the steam engine by Savery, Newcomen, and Watt. In addition, the way to eliminate the sulfur content of coal by the coking process was discovered, and through the use of coke, better steels could be produced. These, in turn, could be used to build better machines.[5]

An interest in such topics led a group of men to meet regularly on the Monday of each month nearest to the time of a full moon. The Lunar Society of Birmingham consisted of only fourteen members. Although some of the members had met earlier, the active years of the society date from 1775 to near the end of the century. James Watt and Josiah Wedgwood were members; Benjamin Franklin was personally acquainted with seven of the members. The society undertook and encouraged scientific experimenting and industrial improvements; its influence was great, far beyond its small membership.[6]

Although better steel could permit the construction of vastly improved machines, the design of better steam engines required a scientific understanding of the thermal properties of steam: The energy available when steam is allowed to expand, the efficiency of an engine and how it could be improved. The science of heat was developing, but was still inadequate.

Thermodynamics was founded when Carnot published, in Paris in 1824, his book *Reflections on the Motive Power of Fire and on Machines Fitted to Develop That Power*, although almost no scientific notice was taken of it until twenty years later and then only in a paper by Clapeyron, another French scientist.[7]

There are, of course, far more details in the development of the theory of heat and the improvement of industrial machines during this period. Some men, such as Joule, were stimulated in their own careers while others, of course, seem to have taken little notice of the revolutionary changes in progress. It is amusing to note the sole comment on this topic in the voluminous diaries of James Boswell; on March 22, 1776, he visited Matthew Boulton, who collaborated with Watt:

> and saw Mr. Boulton's manufactory about two miles from the town . . . I regretted that I did not know mechanics well enough to comprehend the description of a machine lately invented by him, which he took great pains to show me.[8]

Perhaps it is not fair to expect Boswell to appreciate the importance of what was happening at that time, and even with the greater general awareness today there are many who seem oblivious to the technological forces that are being used to alter the shape of our existence and society. (After all, how many of today's automobile drivers know anything of the working of the internal combustion engine?) Certainly from our present vantage point, we are in a good position to see the way in which science and technology were interwoven two hundred years ago, and perhaps we can draw some lessons from it. There clearly was a great need to develop powerful machines, yet this could not be pushed beyond a certain point until the theoretical understanding of heat engines caught up. Within such a climate, the choice of a suitable topic for scientific study can no longer be completely accidental, and Joule's interest is understandable. While this particular episode is seen more clearly because it is less complex than today's situations, we can at least try to see whether there are similar patterns.

7.2   Relevance Today

Our historical digression has been useful in covering some of the points that need to be considered when we discuss the relation of science and technology to society today. There are many other aspects, however, and the study of the impact of technology on a society requires discussion not only of the inventions and industries but also of the social forces and traditions that can either welcome or hinder technological change. Such discussion would divert us from the main topic at hand—the definition of contemporary relevance.

In retrospect we can see the parallel advances of science and technology as improved machines and apparatus permit better experimenting, from which has come a firmer foundation of basic scientific knowledge. In turn, this knowledge can lead to better machines, and so the cycle can repeat. But retrospective recognition is very different from prospective, and science is now confronted with the need to answer a succession of very pointed questions. Is it easy or possible to decide whether some particular research is pure (basic) or applied, and thus qualifies for a higher or lower priority for continued support? In view of the serious problems that advanced societies now face as a result of their "progress," and the problems now facing developing societies, which need locally adapted agricultural and medical technologies, can we afford to subsidise basic science? Should we stop all basic science for a number of years in order to concentrate our efforts and skilled manpower on the immediate and socially pressing problems?

If one should wish to give preferential support to applied research, then decisions must be made as to whether some project is more applied than basic. Some understanding is also needed of the way in which applied research finds its way to the finished product. Just such a study was undertaken a few years ago in the Department of Defense. Project Hindsight was a study of the role of research in the development of some weapons systems during the period 1945–62. As described in a review article, the need for evaluation of the role of research was felt:[9]

> No matter how much science and technology may add to the quality of life, no matter how brilliant and meritorious are its practitioners, and no matter how many individual results that have been of social and economic significance are pointed to with pride, the fact remains that public support of the overall enterprise on the present scale eventually demands satisfactory economic measures of benefit. The question is not whether such measures should be made, it is only how to make them.

To this end,

> one of the objects of Project Hindsight was . . . to try to measure the payoff to Defense of its own investments in science and technology. A second object was to see whether there were some patterns of management that led more frequently to usable results and that might therefore suggest ways in which the management of research could be improved.

Twenty weapons systems were studied. For each the development was traced backward with key contributions being identified until some specific time and place of origin was located. (Sometimes it turned out that two or three sources could be identified, often related.) Such sources were labeled as "Events," and were classified according to the intention that motivated the original work. Classifications used were undirected science, applied or directed science, and technology. The study showed that 91 percent of the Events were classified as technology, meaning that the research had been directed towards the specified end product; 8.7 percent of the Events fell into the applied science category, where the knowledge sought was basic but the end in view was clearly applied. Only 0.3 percent of the contributing Events fell into the undirected science class. The average time taken to develop an improved system was found to be 13 years, and the final product was usually the result of many innovations, the majority of them being "quite modest." The conclusion reached in this study was that[10]

> had the Defense Department merely waited passively for the non-defense sectors of the economy or government to produce the science and technology it needed, our military equipment would be far inferior to what it is today.

The study specifically excluded Events further back than 20 years, and it was recognised that this produced some bias, although it was felt that the time scale chosen was not unreasonable for expected returns on military-oriented research. It was recognized, though, that basic discoveries, many much older than 50 years, were essential in laying the foundations of modern applied science, but it was felt that the applications came less from new knowledge in science than from the well-organized and well-understood "old science."

Project Hindsight attracted much comment and criticism at the time. The Department of Defense was supporting much basic science, and many felt, rightly or wrongly, that this support would change if the Department perceived that its needs would be better served by concentrating on applied research. There was also basic disagreement with the premises and findings of Hindsight. By its nature Hindsight was confined to weapons systems (missiles, torpedos, mines, nuclear warheads) which had had development times of less than 20 years, and the longer-range contributions were excluded. To what extent then can Hindsight be taken as a model for civilian applications? It is worth noting, in passing, that Defense support of basic research is being drastically reduced, starting with the 1969–70 federal budget through the enforcement of the Mansfield Amendment, which requires the military research offices to support only that work which they deem directly related to their immediate objectives. Although this strict provision has been slightly relaxed in the 1970–71 budget, the effect will certainly continue for some time.

To return to the question as to the role of science in technological progress: What can society reasonably expect from science by way of prompt returns, and how can the dividends be increased? The Defense study, Hindsight, seemed to suggest that specific problems are best answered by research which is clearly aimed and that undirected scientific research made very few contributions, at least in those cases studied which could reasonably be taken as representative of the military systems.

This analysis of research narrowly aimed towards a specified goal, omits completely the role of chance in scientific discovery. True, a specified problem may be solved by a carefully planned attack, but what of the problems that have not been singled out for such concentrated attention? We should also note that simply specifying a problem does not mean that we have yet the knowledge or tools to solve it. We need only think of medical problems that have not yet yielded even to heavily supported research; e.g., cancer and multiple sclerosis.

The subject of the element of chance in scientific discovery has been treated by Taton, who cites many cases most of them historical.[11] A more recent survey is contained in a report prepared in 1967 by a special committee of the National Academy of Sciences for the House of Representatives Committee on Science and Astronautics.[12] It contains a series of papers covering all aspects of the applied sciences from agriculture to medicine, the behavioral sciences, industry, and the role of national institutes. In particular, one paper describes the case histories of the development of ten products in the General Electric Company, and these show clearly the roles of chance and direction in the cases described.

The literature on the applications of science is broad, and some references have been included in the bibliography. Although it would be instructive to pursue a few more case studies, space does not permit that here. Instead we can summarize some of the more important points that can be extracted and then turn to some related problems.

Systematic work on a defined problem will often yield results. In so doing, contributions to the solution will be made from established scientific results. Sometimes values of crucial quantities may not be known well enough (if at all), and some basic work may be needed to determine these values. (For instance, in the use of radiation for diagnostic or therapeutic purposes, one needs to know how radiation energy is absorbed in tissue, and so some basic radiation physics is required for the solution of a medical problem.) But chance discoveries also play a role, and we should ensure that these possibilities are not ruled out, as by a rigid requirement of apparent "relevance" in all work.

The difficulty of defining basic and applied science has been repeatedly stressed, as by Brooks who cites an example:[13]

Almost all of Pasteur's work, from the fermentation of beet sugar and the disease of silkworms to the anthrax disease of sheep and the cure of rabies, was on quite practical problems; yet it led to the formulation of new biological principles and the destruction of false ones, which revolutionized the conceptual structure of biology.

There are many other examples that can be cited to illustrate the same point, which can be clearly stated: Even the most obvious of applied research may require the prompt solution of a basic problem, while basic research with no apparent use may well prove crucial for some application tomorrow. About the only thing that can be said with confidence is that there is no general rule to cover all cases. Although in retrospect we can often see the connections between discoveries and research and their applications, the connections or implications may not be at all obvious at the time, and the distinction between basic and relevant research is often almost impossibly hard to make.

It is unrealistic to expect that avowedly basic research work will be supported to the exclusion of clearly applied work. Research is supported by society (through Congress and other bodies) for its anticipated returns. If we could definitely guarantee that there would be no useful returns, then science would probably attract about as much public support as symphony orchestras or archeology. This is, in fact, already well recognized and is clearly reflected in the "R and D" budget figures so often thrown about. The "R," broadly representing basic research, is considerably less than the "D," representing development. What then are the pressing problems of today that require scientific and/or technological help via the "D" category? What basic science can be seen as needed in support of this "D," and then what further basic science can we afford, and where shall it be carried out?

Without trying to set them in any order of importance, we can list the problems that are receiving most attention at present: Environmental problems, which include pollution and urban planning; medical problems; transportation of all kinds, from the development of the SST to rapid urban transit systems; the production of electric power together with the question of fuels needed and, of course, the resulting pollution; technical assistance to developing countries; and population growth. Military research and development should also be listed, not necessarily for its popularity but because it still absorbs large amounts of money.

In each of these areas and others that can be added according to personal preferences, much progress can probably be made simply by the persistent application of current scientific knowledge. It can also be said that most of these areas require attendant basic research, which will either considerably expedite solution of the main problems or from which there may come those lucky accidents that will be seized upon by a waiting but prepared scientist. We can answer the question "Can we afford to subsidize basic science?" by saying that we cannot afford to neglect basic science.

NOTES AND REFERENCES

1. Lynn White, Jr., *Scientific American*, **223** (August, 1970), 92.

2. Edward MacCurdy, ed., *The Notebooks of Leonardo Da Vinci* (New York: George Braziller, 1956), 806.

3. *Ibid.*, 850.

4. Galileo Galilei, *Dialogs Concerning Two New Sciences*, transl. by Henry Crew and Alfonso De Salvio (New York: Dover Publications, Inc., n.d.). Reprinted through permission of the publisher.

5. For a more detailed treatment, see, for example, Charles Singer, E. J. Holmyard, A. R. Hall, and Trevor I. Williams, eds.: *A History of Technology, Vol. IV: The Industrial Revolution* (London: Oxford University Press, 1958).

6. Robert E. Schofield, *The Lunar Society of Birmingham* (London: Oxford University Press, 1963).

7. Sadi Carnot, *Reflections on the Motive Power of Fire* (New York: Dover Publications, Inc., 1960).

8. Charles Ryskamp and Frederick A. Pottle, eds., *Boswell: The Ominous Years* (New York: McGraw-Hill Publishing Co., 1963), 289.

9. Chalmers W. Sherwin and Raymond S. Isenson, *Science,* **156** (1967), 1571. (Copyright 1967 by the AAAS.)

10. *Ibid.*

11. R. Taton, *Reason and Chance in Scientific Discovery* (New York: Science Editions, Inc., 1962).

12. *Applied Science and Technological Progress: A Report to the Committee on Science and Astronautics, U.S. House of Representatives, by the National Academy of Sciences* (Washington, D.C.: U.S. Government Printing Office, 1967).

13. Harvey Brooks, *Science,* **156** (1967), 1706. (Copyright 1967 by the AAAS.)

science and government

chapter 8

The facets displayed by the contacts between science and government are countless; some can be discerned in historic episodes, others are quite recent creations. The literature in this field, even in dealing with individual cases or problems, is immense. Accordingly a relatively modest goal has been set for this chapter. By frankly personal choice, a selection of topics will be treated briefly to indicate some of the more important issues involved. The bibliography provided is extensive enough to allow the reader to continue his study into whichever aspect may excite his interest.

8.1 Scientists As Advisers

In an earlier chapter, reference was made to military-related research. Scientific advice has long been sought for the improvement of military systems and defenses. While today advice can be obtained from scientists who are employed in one or another of the various government departments or agencies, much use is still made of advisory panels which include university and industrial scientists. Major areas of decisions involving such panels have covered nuclear weapons, civil defense, chemical and biological warfare, and antiballistic missiles. At a much lower level would be decisions on technical problems such as might arise in communications, night vision devices (using infrared sensors), or a host of other military details.

Decisions on major problems are no longer the sole interest of the

Department of Defense. The scale of costs clearly interests Congress, while international implications just as clearly interest the President and Department of State. For scientific advice the President can call upon the President's Science Adviser, and the President's Science Advisory Committee (PSAC), which is composed of Administration appointees. Whereas originally PSAC was primarily called upon for advice in military matters, it has increasingly been consulted for help on other topics, ranging as far as world food problems and the protection of humans who are the subjects of research in the social or life sciences.

This advisory role is a channel through which the scientific community can try to exert its influence. Even a casual inspection of the record, as can be obtained by leafing through the topical news reports in *Science* for the past few years, will show a very mixed collection of successes and failures. Among the latter one can note that despite a continuing barrage of testimony the total federal support for basic science is slowly dwindling. Sometimes advice may be given via PSAC, but at other times it appears at Congressional hearings, especially in matters involving technology. Witness the recent, fierce debate on the merits of continuing the development of the supersonic airplane. By comparison, despite an overwhelming preponderance of scientific advice against the building of an ABM system (in 1969), Congress voted approval and funds.

What we see in all of this is a call for scientific advice, with no guarantee that it will be followed. In some issues one might consider that the weight of expert advice should be a strong persuader when the problem is primarily technical. Experience shows, however, that once a political, diplomatic, or important economic side is seen, scientific opinion alone can no longer be expected to carry the day. Many of these issues have aroused violently partisan proponents and opponents, which places a particularly heavy burden on scientists who bring expert advice. They must do all they can to separate their expertise from their politics so that the different components of a final decision can be clearly distinguished and combined with appropriate relative weight. Unfortunately this has not always been done, to the resulting public confusion and also to the detriment of the reputation of scientific objectivity.

While the Administration will call in the PSAC to help grapple with cosmic issues (or at least those involving a lot of money), scientific advice is also sought at virtually every lower level of federal, state, and local government. It is easy to overlook the immense quantity of expert advice that is continually being evaluated. The Federal Drug Administration is responsible for the evaluation of tests on all new drugs before they can be licensed for general use. Agricultural and domestic pesticides and herbicides must similarly be scrutinized, and decisions must be made on whether these are safe for general use or only for restricted applications and then on what warnings need to be printed on the package labels. The National Bureau of Standards carries out tests and sets standards. As with military technology, decisions in these other areas will be based on expert advice both from laboratories within the departments or agencies as well as from outside consultants.

When we talk of scientific relevance, then we should include the work of laboratories, panels, and agencies such as we have mentioned. There is almost no part of life in a modern industrial society that does not depend on the use of applied science; the corollary is that scientists and technologists will be involved in drawing up safe standards and in implementing and reviewing these standards in addition to processes and products. There has been much criticism recently of the shortcomings of the system or of individual agencies, but it would seem that there can be no escape from an even greater dependence on expert advice and review.

It is particularly in the great public debate on environmental problems that we are seeing pressure for the extension or adaptation of some part of the federal regulatory system to the state or local level. Despite resistance mostly industrial, there is widespread agreement that standards must be set for the cleanliness of our air and waters. Who will set these standards? What represents a reasonable balance between the expedient and the desirable, between the practicable and prompt, and between the desirable but politically unlikely?

There is a great deal of evidence on the deleterious effects of certain pollutants but not much information on others. In addition, pollution is a combination of effluents from many processes and plants. Jurisdictional disputes arise. For instance, when a utility has received approval from the Atomic Energy Commission (AEC) for the construction of a nuclear reactor for the generating of electricity, the AEC requires that certain standards be met with regard to the release of any radiation. Does the state where the plant will be have the right to set stricter standards than those of the AEC for emission of radioactive products? Such a case is now being argued in the courts for a proposed nuclear reactor in Minnesota. In addition, nuclear (and other) plants generating electricity require large amounts of cooling water. If such a plant is located at a lake shore, large quantities of heated water will be discharged into the lake. Who shall decide what temperature of emitted water is safe with due regard to the plant and animal life in the lake? The AEC may license the plant, but until very recently the AEC did not review the hot water and environmental problems, and the jurisdiction is not yet well defined.

In these and thousands of other environmental problems, scientific advice is being sought. Some states and local authorities have set up advisory groups.[1] Some of these groups have full-time staff, some part-time. The problems are not entirely scientific, but have legal and political ingredients. Often neighboring states will cooperate, but not always. This makes effective regulation impossible, as for instance in setting clean air standards where industrial smoke can be blown across a state boundary.

Involvement in environmental problems places the scientist in a position rather different from his usual professional stance. As has been discussed in Chap. 3 and as Ziman has so well described, science advances by consensus. There is no formal vote as to whether or not to accept some new result; it is either compelling in itself—or it is not. In the latter case, the scientists can reserve judgment. They may feel that the result is correct or not and use their judgment in planning further experiments that will provide sharper tests, but no final commitment is needed. Advice in the public domain, especially when public health is concerned, requires different weighing. For instance, it may be suspected that a particular food additive can cause birth defects. The experimental evidence may be based on a small number of laboratory animals, and even though this additive may have been in use in foods for several years, clinical studies may not yet exist to show how it affects humans. Should this additive be immediately withdrawn or banned, or can its continued use be permitted pending further tests and review? This is not completely hypothetical, for the biological effects of food additives, pesticides, fertilizers, and many other chemicals are now being queried. The decisions must often be made on less than complete scientific data, and other factors must obviously enter into the judgment. The scientists on advisory panels will be called upon to recommend action and will often not be able to afford to take a conservative wait-and-see attitude.

Decisions such as these involve scientists in ways that are not usually visible to the public. Sometimes the decision involves the product of a major company, which may try to exert pressure for a favorable ruling, and it has happened that word of such pressure reaches the press, but more usually regulation in the public interest arouses the attention of only those professionally concerned.

There is a very different arena in which scientific advice is sought, and where the glare of publicity is deliberately used by competing parties. The scale of "big science" is now so large that there is fierce competition for each new facility. Here we include the siting of a large new accelerator for fundamental nuclear physics experimenting or the siting of a new major national laboratory, such as that to handle samples returned from the moon. There is much more at stake than just the laboratory named. That alone would create hundreds of new jobs in the chosen region and can involve millions of dollars each year through salaries and local spending. There is also considerable prestige, and local universities may try to build research groups around the new facilities and thus be more attractive to the faculty and students they wish to draw. Local chambers of commerce recognize that these facilities can help attract new technological industry; the net result is a concerted effort to gain the award of large scientific complexes. When the new national accelerator (near Chicago) was being planned, many states and cities competed, and the state of Colorado and the University of Colorado (in Boulder) submitted their proposal in a handsomely prepared booklet complete with a full-color photograph of snowy scenery on the front cover.

Scientists will clearly be on the panels that consider all prospective sites and will be trying to assess the factors for and against each one: Ease of access for heavy equipment as well as for scientists who will commute from many universities, proximity to good universities and local technical industry, adequacy of local electric power supplies, and many other factors. Scientists who have a major interest in the outcome will not be on a panel but will be competing against one another for a favorable decision—again this is very far from the normal conduct of science.

The role of scientists in decision making has been discussed extensively by many authors,[2] and the professional attention of political scientists and sociologists has been attracted. When power and influence are at stake, the scientists have shown themselves to be refreshingly human and fallible, which might be a source of comfort for those who have feared a complete takeover by robotlike scientists and technologists.

Government at all levels will inevitably require scientific and technical advice, and it is not surprising that there should have already occurred situations in which scientists have clashed with the authorities (both temporal and spiritual). Such disagreements can pose severe threats to the intellectual independence of the scientific community. Some specific cases where such clashes have occurred will be reviewed in the remainder of this chapter. The first case to be examined will be that of Galileo. It might be thought that his case represents a form of conflict that will no longer arise, in that the issue was between a scientist and the Church, whereas today we are mostly concerned with disagreements between scientists and their state. It is, therefore, worth pointing out that there is still a major problem today which involves the Church, and that is population control. In the concern over the progressive degradation of the quality of our environment and the problems associated with providing adequate health care and food to the population, much attention has been focused on the rapid rise in world population, and especially in certain countries. Any program for population control must be based on scientific knowledge—information from the biological sciences regarding the various methods of contraception and their side effects, information from the social sciences regarding different approaches and methods for different countries and cultures—but with obdurate Church opposition, a clash would appear inevitable. Caught in the ideological middle will be those scientists whose professional advice runs counter to the proclaimed view of their own Church.

8.2   Galileo—and the Freedom to Disagree

Galileo occupies a unique position in the history of science. He was the first clearly to understand the role of controlled experiments, and his methods of reasoning represent a sharp break with earlier scientists. Modern dynamics, involving concepts such as inertia and velocity starts with Galileo. Newton later made these concepts quantitative where Galileo had been more qualitative and then went on to describe gravitational effects. Galileo, however, laid much of the foundations and was concerned with the science of mechanics for much of his working life from his unpublished manuscript "On Motion" (c. 1590) to his great treatise *Dialogues Concerning Two New Sciences*, published in 1638 only four years before his death.

His other great interest lay in the area of astronomy, and here again his contributions were monumental. He has described in his *Starry Messenger* (1610) how he made his own telescope, the first to be used for astronomical purposes. There followed the first observations of the four major moons of Jupiter, the phases of the planet Venus, and the observations of spots on the sun.[3]

Galileo had long believed in the heliocentric model of the solar system as described by Copernicus and had expressed this belief in a letter to Kepler as early as 1597. His new observations (especially those of Venus) strengthened this belief.[4] His treatise *Dialogue Concerning the Two Chief World Systems* (1632)[5] explored the merits of the systems in which either the Earth or the sun is at the center of the solar system, and there is no doubt which he believed to be true. In addition, the discovery of Jupiter's moons provided Kepler with a test of his newly deduced laws of planetary motion. Kepler was able to borrow one of Galileo's telescopes (through an intermediary), repeat and confirm Galileo's observations of the moons of Jupiter, and show that the same laws held for Jupiter and its moons as held for the sun and its planets. (Despite Kepler's demonstration that the planetary orbits were elliptical, Galileo persisted in his belief that they were circular.)

Although the scientific history of this period is fascinating, its technical interest will probably be limited to those with some understanding of physics and astronomy. It is the nonscientific aspects of associated happenings that has attracted widespread attention.

At that time the Church maintained that the heavens were unchanging and that the Earth was the center of the universe. These dogmas reflected biblical interpretation, and the Church was in no mood to consider yielding while it faced challenges from the rising Protestant faith. There had, of course, been innumerable observations of transient phenomena in the skies, even without the aid of the telescope—comets, meteorites, and occasional exploding stars (novae and supernovae). Some of these are very well documented. Tycho Brahe, in 1572, recorded a supernova, seen as the sudden extreme brightening of a star, with the increased brightness continuing for months. Kepler in 1604 observed something similar. The Crab nebula in the constellation Taurus is the remains of the supernova of 1054, which was initially so bright that it could be seen by day and was even brighter than Venus. To all of these observations, the Church remained silent, possibly because none was tied to a heretical theory but remained isolated sightings.

Galileo's observations with the telescope presented a more direct threat. But as Galileo was a devout member of the Church, he was almost certainly more interested in seeing the Church support the truth than in seeing the Church confronted. As reported in his books, other celestial bodies than the Earth were seen to have satellites, and changes did indeed take place in the heavens since spots were seen to occur and disappear on the face of the sun.

These discoveries and their descriptions attracted much attention and hostility, and they were the subject of violent attacks from the pulpit as well as in printed pamphlets. Galileo's reply appeared in 1615 in his "Letter to the Grand Duchess Christina."[6] He was summoned by Cardinal Bellarmine and appeared before him on February 26, 1616, and agreed that in future he would neither hold nor defend the Copernican views. When his *Dialogues Concerning the Two Chief World Systems* appeared in 1632, it was clear that he still held to his views, and he was called before the Inquisition, where his agreement of 1616 was brought forward. But where Galileo maintained that he had agreed not to hold or defend those views, the Holy Office produced minutes of the 1616 meeting which went further, enjoining him from "teaching, in any way." These minutes of the 1616 meeting are unsigned— remarkable for so meticulous an organization as the Inquisition. In his own defense, Galileo was able to produce a letter from Cardinal Bellarmine to him (also of 1616) which agreed with his own recollections. All of these documents are still available to scholars, but the authenticity of the 1616 minutes has become the center of scholarly disagreement.[7]

In any event Galileo finally signed an admission that conceded he might indeed have been warned as alleged, but that he might have forgotten the additional terms. He was sentenced to a form of house arrest, and it was in these final years that he returned to his interest in mechanics and completed his *Two New Sciences*.

What is involved is more than simply the possible defiance of an order from the Church. There was fierce competition between the Jesuits and the Dominicans for power within the Church; it has been suggested that Pope Urban VIII needed to divert attention from defeats suffered during the Thirty Years War. Although all of this may seem far removed from our modern concerns of science and government, it remains an episode with current analogies.

8.3   The Oppenheimer Case[8]

Despite the gap of more than 300 years that separates this case from that of
Galileo, there are sufficient parallels to make their comparison instructive.
In both we are dealing with strong-willed individuals of very great ability,
whose views ran contrary to that of established power. In both cases the
established authority was engaged with external problems so that it might
seem a sign of weakness to permit too much latitude to prominent dissent
at home: The Pope was much concerned with alliances in the conflicts across
Europe; during the mid-1950's, the cold war and the McCarthy era at home
led to considerable intolerance of dissenting views. But any analogy can be
pursued too far, and the comparison here is of use only in guiding our
attention in the examination of the more recent case.

Where Galileo made major original contributions to science, Oppenheimer's contributions were at a lower level of originality and are quite
overshadowed by those of many of his contemporaries. It was as a leader
and stimulating teacher who could inspire others that he is best remembered,
apart from his involvement in the Manhattan Project.

The story of the development of the atomic bombs in World War II
has been told many times, and books on the Manhattan Project and its
participants including Oppenheimer are increasing. Although the events have
been described by some of the participants, most of the newer literature is
coming from nonscientists, who are drawing different comparisons and
making different judgments.

With the use of the two bombs over Japan and the end of the war in
1945, the fate of further nuclear research and development in the United
States had to be considered. The first major decision, made in 1946, led to
the formation of the Atomic Energy Commission (AEC), through which all
work on atomic bombs would be centralized. This was set up as a civilian
agency of the Administration, the defeat of the May-Johnson bill in Congress
ensuring that nuclear energy would not be controlled by the military. The
May-Johnson bill had been bitterly opposed by many major scientists,
which is an interesting story in itself, but it is of interest to recall that Oppenheimer supported the adoption of the May-Johnson bill.

The continued development of atomic weapons was then placed under
the control of the AEC, with the cooperation of the Department of Defense,
and the strict wartime security precautions were continued. Those working
on this and other sensitive projects had to obtain "security clearance,"
which entailed detailed investigation of political associations and background. With the growing cold war and the obsession with atomic secrets
finding their way to Russia, the McCarthy era was starting. Many scientists
had liberal political views, and some had had close associations with Com-

munists before the war. To an often uneasy feeling as to the reliability of scientists in general (too often caricatured as foreigners in white coats, with beards, thick glasses, and equally thick accents), there was added the spy scares of the early 1950's. Two scientists were jailed in England for passing information to Russia; both had had associations with the wartime atomic bomb project on this side of the Atlantic. Suspected spies were arrested in this country; two were executed after being found guilty and others were jailed. The evidence that formed the basis for the successful prosecutions has been challenged in the last year or two, but that does not affect the mood which was then widespread in the country, especially after a presidential election (1952) in which one of the major slogans had been "twenty years of treason."

Oppenheimer had been the director of the Los Alamos Laboratory during the war and had directed the scientific aspects of the weapons development and construction there. When the AEC was established by law in 1946, a General Advisory Committee (GAC) was also set up to provide the AEC with scientific advice. (The Commissioners have usually been nonscientists.) Oppenheimer was one of the initial members of the GAC, and was elected chairman, a position he held until his resignation late in 1952 shortly before the presidential election. During this time one of the major decisions that the GAC had to consider was the advisability of the AEC's promptly starting a crash program for the design and construction of the H-bomb. (Until that time all atomic bombs had obtained their energy from the fission or breaking apart of heavy atoms such as uranium; the H-bomb or hydrogen bomb is based on the release of energy when hydrogen atoms are fused into helium, a heavier atom. This fusion process, which requires higher temperatures for its initiation than for the fission process, permits far greater explosive power to be obtained.)

Within the GAC, after intensive discussion, the decision went against recommending the implementation of a crash program. Oppenheimer was part of the majority in this vote. Perhaps the GAC might have decided differently if it had been asked about a slower program, but in any event it advised against the crash program. However, when the Truman Administration decided to push ahead with the H-bomb project in 1950, the GAC administered the program and supported it.

In August 1953 the first Russian H-bomb was detonated, considerably sooner than had generally been expected, less than a year after the United States had had its own first H-bomb. This aroused in many quarters concern for the adequacy of the United States' arsenal, as well as questions concerning the vigor and secrecy of the weapons program in this country.

There is still dispute as to the exact sequence of and motives behind the moves made around this time that were aimed at separating Oppenheimer from the weapons program and at preventing him from advising on scientific policy. He was denied access to classified information and was so informed in a letter in December 1953. That letter went on to itemize instances on which the decision was based. Many of the events cited involved Oppenheimer's movements and associations of many years before and had been well-known when he worked at Los Alamos. Certainly at that time they had not been considered serious enough to separate him from the program. But the letter went further:[9]

> It was further reported in the autumn of 1949, and subsequently, you strongly opposed the development of the hydrogen bomb; (1) on moral grounds, (2) by claiming that it was not feasible, (3) by claiming that there were insufficient facilities and scientific personnel to carry on the development and (4) that it was not politically desirable. It was further reported that even after it was determined, as a matter of national policy, to proceed with the development of a hydrogen bomb, you continued to oppose the project and declined to cooperate fully in the project.... It was further reported that you were instrumental in persuading other outstanding scientists not to work on the hydrogen bomb project, and that the opposition to the hydrogen bomb, of which you were the most experienced, most powerful, and most effective member, has definitely slowed down its development.

Faced with the choice of simply accepting this restriction or of going through a formal hearing, Oppenheimer chose the latter. A three-man board heard the case during April–May, 1954, and finally, in a 2–1 decision, recommended that Oppenheimer continue to be denied clearance, but it did not find him guilty of disloyalty. The majority, however, did consider him guilty of poor judgment in several instances and referred to "fundamental defects in his character."

About 40 witnesses were heard by the review panel. Some scientists gave strong testimony in Oppenheimer's favor, whereas a few gave equally strong evidence against. Some of those who had testified against Oppenheimer, notably Edward Teller, were ostracized by other physicists for several years, and even at this date memories have not been erased. The testimony was originally intended to be held confidential, and witnesses were repeatedly reassured of this. Nevertheless, the transcript of the hearings, censored of secret material but still amounting to nearly a thousand pages, was issued by the Government Printing Office that same year (1954).

The case aroused intense feelings at the time, and even now it is still a matter of great interest. Haberer has given it a detailed analysis in his book along with other examples of the contact between science and politics, while Kipphardt[10] has written a play drawing his lines from actual testimony as recorded in the official transcript of the hearing, but Oppenheimer protested the dramatic portrayal's emphasis.

Among the many aspects that present themselves for detailed study, two only will be selected here: First, the role of Oppenheimer as official scientific adviser, and second, the effect of this case on other scientists.

Unless an advisory group maintains strict secrecy about its internal divisions and always presents unanimous reports and recommendations, one would expect to find policy recommendations that sometimes are compromises and sometimes simply majority positions. It is unreasonable to expect genuine unanimity on all issues unless the decisions are so diluted as to evade the real difficulties. Policy will hopefully be best (however that may be defined) if it emerges from honest discussion between different views. Furthermore, unless different viewpoints are tolerated, one is likely to end up with unanimous but possibly unwise or wrong decisions. Full discussion and diversity of opinion will not be found on advisory panels if holders of unpopular or minority views are subject to prosecution. When a scientist disagrees fundamentally with a policy he is called to supervise, what should be his move? It has been suggested that Oppenheimer should have resigned when Truman decided to proceed with the H-bomb development. In fact, he considered doing this but was dissuaded from such action by Gordon Dean, then chairman of the AEC.

Alternative to resigning is to remain as adviser to ensure that opposition opinion continues to be represented with the possibility that subsequent policy may be thus influenced. This role of "loyal opposition" may be quite unfruitful (depending on circumstances), and it is a matter for personal opinion and critical timing to decide when one's purposes are best fostered by remaining on a committee or resigning in public protest.

It would probably be generally accepted that committees which set or recommend political policy should reflect the views and philosophy of the administration that appoints them. What is not so clear is the need for political conformity on advisory panels that are intended to be primarily scientific. In addition to the many other sides to this case, what is being seen is an inherent conflict between two tasks assigned to the AEC: The design and production of weapons on the one hand, and control and protection on the other. This conflict of AEC roles emerged again some years later when the controversy arose regarding the biological effects of radioactive fallout from weapons tests, and in fact this particular subject continues to be the center of argument. Political conformity in administration appointments has also continued to be a problem and will be treated again later in this chapter.

The second main point to be considered is the reaction of other scientists. As has been well documented by Haberer, organized or public expression of opinion on this case essentially ceased at the end of 1954. There was very strong expression of views at the time of the hearings, and the *Bulletin of the Atomic Scientists'* pages contain much protest against the AEC, and against the security system from which this case had emerged.[11] Other journals gave far less coverage. Haberer criticizes the scientific community for not having given greater support to Oppenheimer and for so quickly losing public interest. A quick judgment on the merits of this criticism would be unfair, and this question is left as an exercise for the reader. What needs to be taken into account in reaching a judgment are factors such as the avenues of protest available—to organized science and to individuals—the improving political climate, especially following the Senate censure of McCarthy in December 1954, and the emergence of other scientific-political issues such as that pertaining to radioactive fallout. In fact, as will be detailed in Chap. 10, scientists were increasingly outspoken.

It is not clear that the hearings and the decision reached resolved any issues other than the immediate one of continued clearance for one individual. The proceedings must be viewed against the political picture of that time. Some official recognition of error may be inferred from the decision of President Kennedy to nominate Oppenheimer for the Fermi Award in 1963. This award, for services in the cause of atomic energy, was opposed by some who had opposed Oppenheimer in 1954, and even with the award his security clearance was not automatically reinstated; rather, the AEC took the position that he would need to reapply if he wished to have a clearance again, and this Oppenheimer did not do.

There are certainly morals to be drawn from a reading of the documents that cover this affair. Each will have to draw his own conclusions, and it is not unlikely that still-secret information may emerge to make us alter our opinions. In the confusion of good intentions, foolish words and actions, and questionable motives, we can once again observe (and be reassured by?) the very human responses of the participants, quite apart from whether we agree with them or not.

8.4   Scientists As Advisers: Patronage or Independence?

The Oppenheimer case raised the problem of the independence of scientific advisers. At the highest level, it was made to seem that more than loyalty was demanded, even though other, more personal factors probably also played roles. For an administration to be sure that its policies were being vigorously implemented, it would seem reasonable for appointees to sensitive policy-making bodies to reflect the views of that administration. Over the years, this problem has generally been avoided, as scientists strongly opposed to an administration or to particular programs have tended to keep out of the way of any close involvement. For instance, it seems as though there must be very few who work on the nuclear weapons program who disagree strongly with it. The opposition has come primarily from those outside, working mainly in the universities. If they are not privy to secret information, they are more free to voice their criticism, but by the same token they are often accused of not having all the information available to those who do have clearances and thus access to secret information. When a scientist accepts appointment to a panel and with it the access to classified information, he may also be accepting some limitation on his freedom to criticize.

The ABM controversy was therefore unusual in that the most persistent and forceful arguments against the program came from scientists of national stature, scientists who had been advisers to all previous presidents. The arguments they presented publicly were, of course, not classified, but they were also able to assert from their knowledge of the secret information that their arguments would not be changed even if what was secret were to be made public. Many other scientists also have argued strongly against an ABM system of any kind, and in favor of arms control. One of these who has worked vigorously for control has been Franklin A. Long, vice president for research and advanced studies at Cornell University.

When the Nixon Administration took office in 1969, one of its early tasks was to search for a new director for the National Science Foundation (NSF). The NSF, operating with an annual budget of close to $500 million, is the prime federal agency for the support of basic science other than in the medical areas (which are strongly supported by the National Institutes of Health). Long was considered to succeed Leland Haworth at the NSF despite his support for the Democratic party through the scientists' efforts for Johnson in 1964 and Humphrey in 1968. His nomination was approved by the National Science Board, which sets policy for the NSF and the prospective appointment was also approved by members of the congressional delegation from New York. An appointment was set for Long to meet with the President on April 11, 1969, after which the formal announcement would be made, and press releases were prepared at both the NSF and Cornell. When Long arrived in Washington around midday on April 11, he was told that "The situation had changed and that the new elements of a political nature relating to the antiballistic missile system had arisen." The appointment to the NSF had been withdrawn.[12] News reports present a rather more confused picture, and it seemed that objections had been made against the appointment by a member of the House Committee on Science and Astronautics.

There was strong criticism in the scientific community, and many pointed out that it would be hard to get competent men to agree to undertake major tasks if they were to be exposed to tactics that might be normal procedure in purely political matters but that should have no place in science. In the light of this strong expression of views, at a meeting of the National Science Board on April 28, the President announced that the next NSF director would be chosen solely on the basis of administrative and scientific competence, and that the Administration had been wrong to act against the appointment of Long.[13]

At no stage was organized science involved in the selection of a prospective director, but rather the Science Board, through its contacts in the scientific community, considered several prospects from whom a selection was made. Most scientists had considered implicitly that the NSF was removed from party politics, and the maneuvering in this case came as a great shock.

In another area of science advising, screening procedures for prospective advisers were found whose existence had not been generally known. The Department of Health, Education, and Welfare (HEW) has numerous panels to review the applications for research funds in the fields of the medical and biological sciences. These are expert review panels whose very considerable skill and experience are in large measure responsible for the high level of the programs chosen for support. It turned out, however, that a number of scientists, some even of great eminence, including one who would shortly win the Nobel Prize, were excluded from participation on panels simply for political reasons. In no case whatever were these panels called on to deal with secret research. Lists of ineligible scientists were drawn up, one estimate being that 200 such scientists were so listed. To be known to have been thus blacklisted could well work against a scientist, and there was very strong protest when the existence of blacklists became publicly known. *Science* during the second half of 1969 gave detailed coverage to this relic of the McCarthy era, and under pressure HEW agreed to discontinue the use of this system.[14]

The general point must still be met. Can science escape politicization when it seems as if the tendency is towards taking sides in so many other aspects of daily life? Many scientists still hope that in purely professional matters, it should be only a matter of individual competence that counts. The problem cannot be thus avoided, however, for then it hinges on the definition of the term "purely professional" or "purely scientific."

The question might then be phrased somewhat differently. Should Federal agencies that deal with science, such as the NSF or NIH, be elevated to cabinet level and their heads made clearly political, changing with each administration? Or should the appointments be quasi-civil service as the AEC has become? In addition, it has been argued that a governmental scientific adviser would have far greater influence if he were a member of the cabinet. Alternatively, should science be divorced from politics as far as possible? If not, might not one expect to find republican and democratic candidates for election to (for example) the presidency of the American Chemical Society? Although questions of this kind have been asked sporadically, generally no acceptable answers have yet resulted; while Congress and various administrations grope their way between conflicting interest groups, it is not surprising that science occasionally gets bruised in the process.

8.5   The Interference of Government in Science

In the case of Galileo, scientific ideas themselves confronted the Church doctrine. When restrictions were placed on Galileo, the intention was to prevent the further dissemination of ideas that would undermine the accepted teachings, but Galileo was still permitted to continue his work in isolation. With the other more recent cases that have been examined, the disagreements between scientists and government have not involved the validity of scientific ideas but rather the advocacy of political policies. Oppenheimer, for instance, was denied access to secret information, but no attempt was made to dictate to him which physical theories were correct and which were wrong. Furthermore the AEC made no attempt to evict from science all those who believed in the correctness of whatever theories Oppenheimer would use in his own calculations.

The last two sentences of the above paragraph will probably seem so preposterous that their inclusion might be queried at once. But just such actions or others closely related have indeed been taken against scientists by governments within the past 40 years in other countries. Although the motives behind the actions were primarily political, the result was direct interference in the conduct of science. There have also been instances where scientists were victimized for being outspoken on political matters without the authorities attempting to meddle or dictate directly in the working of science itself. This type of case, which would include the McCarthy period in the United States in the early and middle 1950's, will not be considered here. The two instances that will be dealt with are German science in the 1930's and genetics in Russia during the 1930's and 1940's.

The National Socialist Party came to power in Germany in 1933, with anti-Semitism a prominent part of its mode of action. A succession of laws was aimed at the removal of Jews from all important public positions including university posts. Starting already in 1933, the universities and scientific institutes were purged of almost all Jews, so that many Jews were forced to emigrate to England, Canada, and the United States. Fleming and Bailyn described this movement, which affected not only the sciences but also the creative arts and the professions.[15]

Attempts were also made to discredit the contributions made by Jewish scientists, particularly Einstein. A leading role in this campaign was played by two Nobel Prize winners in physics, Lenard and Stark, who had been supporters of Hitler as far back as the early 1920's. Illustrative of this was some correspondence in the columns of *Nature* early in 1934 between Stark and Professor A. V. Hill, a noted English physiologist.[16] Stark was probably sincere in his warped views, which in that correspondence are mainly concerned with attempting to justify the anti-Semitic discriminatory actions then being taken. His views on scientific matters come across in an article that he wrote for *Nature* in 1938:[17] He tries to draw a distinction between pragmatic scientists, whom he praises for being "directed towards reality," as opposed to scientists of the "dogmatic school." A "dogmatic" scientist he says

> *starts out from ideas that have arisen primarily in his own brain, or from arbitrary definitions of relationships between symbols [and] by logical and mathematical operations he derives results in the form of mathematical formulae ... the dogmatic spirit leads to the crippling of experimental research and to a literature which is as effusive as it is tedious....*

With this background Stark elaborates on his bizarre theme, leading later to the comment,

> *I have also directed my efforts against the damaging influence of Jews in German science, because I regard them as the chief exponents and propagandists of the dogmatic spirit.*

It is interesting to note that this article provoked precisely zero response in correspondence from the readers of *Nature*.

Although some attempts were made to eliminate the teachings of Einstein, competent scientists saw that this was incompatible with any progress in atomic physics. Einstein's special theory of relativity was required for detailed calculations, and Heisenberg, a theoretical physicist, in the end appealed to Himmler, head of the Gestapo, for protection against the attacks that Stark had made on him because of his use of the theory of relativity. Haberer[18] has given a good overview of the effects of the purges on German science, and it seems that the results lay mainly in the direction of victimizing persons while the attempts to lay down scientific doctrine were clumsily and inefficiently handled.

Very different from this was the situation of the science of genetics in Russia during the Stalin years. Here the state interfered in a massive way with the conduct of research and development. One school of thought was strongly supported for around 30 years; scientists who supported the opposing school lost their positions, were jailed, or simply disappeared. It is a complex episode (era might be a better term), and it is only now that many of the details are emerging.[19]

The starting point was the need for vastly expanded agricultural production in the USSR. In 1931 the government published a decree that posed problems to the Lenin All-Union Academy of Agricultural Sciences (LAAAS) and the All-Union Institute of Plant Breeding (AIPB). The problems dealt with developing varieties of cereal crops adapted for different climatic regions; the decree demanded that the ten years normally required for such development be shortened to four years. Other unrealistic demands with similarly short time spans were set. N. I. Vavilov, then president of the LAAAS and the AIPB, realized that these demands could not be met, but

T. D. Lysenko published a pledge to produce the needed varieties in two and a half years.

From this point on, the story is one of increasing favor and honors, for Lysenko, and ignominy and persecution for Vavilov and his followers. Some of this was known in the west and in 1948 and 1949 several articles appeared that described the status of the science of genetics in Russia and pointed out the political aspects.[20] A major source of information has appeared only in 1969 with the publication in the United States of the book by Zhores A. Medvedev, a younger Russian geneticist.[21] This book has not yet been permitted publication within Russia, and the American edition was translated from a microfilm copy of the manuscript that had been sent out of Russia. Apparently parts of the manuscript had been quietly circulating within Russia for several years and had been corrected by its author in the light of comments from those who had been closer to the events described. *The Rise and Fall of T. D. Lysenko* documents, as well as is now possible, this remarkable affair in which Lysenko, a brazen charlatan, was able to dominate the scene so effectively for so long, while brilliant and competent scientists (Vavilov and others) were physically obliterated and with them (for many years) a distinguished and eminent school in Russian science.

The scientific context of the debates, which turned into deadly struggles, centered on genetics. The accepted view in science is that those species survive that are best adapted to their environment, and genetic laws that deal with random appearance of mutations can describe the observations. The names of Darwin and Mendel are associated with the early stages of this science. An early view, long discarded by geneticists, held that adaptation was transmitted by inheritance: If an animal needed to stretch his neck to reach higher leaves, this characteristic would appear in the next generation in the form of longer-necked animals. As an alternative, it was held and considered for some time, but the accumulating evidence on the existence and properties of genes led to this theory's being abandoned. The names of Lamarck and, in Russia, Michurin are most closely connected with this now discarded view, which came to be called "Michurinist."

This case is without parallel in modern science. Dobzhansky, in a symposium on Soviet science in 1951, had this description:[22]

> *Thus far, neither Lysenko nor any one of his numerous followers, have produced a single new or original idea, either a right or a wrong one. It can be stated without hesitation that michurinist biology is nothing more than a relapse towards views that were current in biology in the nineteenth century, and which were discarded early in the present century, mainly owing to the discoveries of genetics. The sum total of what the lysenkoists have to offer is abandonment of the chief attainments of biological research and thought during the current century.*

Others who met Lysenko have also commented on his surprising ignorance of plant physiology and genetics. Lysenko initially worked at the Odessa

Institute of Genetics and Breeding, where he conducted experiments that led him to his theories, or, rather, to the reviving of the Michurin views. His experiments were often attempted by scientists in the west, but his results could not be repeated; in one experiment, described by Medvedev, his radical conclusion on inherited characteristics was reached on the basis of the observation of a single plant, with no controls.

In marked contrast, Vavilov had built up an internationally respected school, and his recognition of scientific methods precluded his making unrealistic claims. Lysenko, on the other hand, went on to make more and more sweeping claims, so that at a 1935 Congress he was applauded by Stalin. As Medvedev has written, "And after Stalin's famous 'Bravo, Lysenko, bravo!' a new and special period began in Lysenko's activities and in the history of Soviet biology."[23]

We need not chronicle here the many further sordid steps; the documentation is now public.[24] Despite continual failure to actually produce the promised results and despite the continued sorry state of Soviet agriculture, Lysenko remained in favor, even when Kruschev came to power after the death of Stalin. It was only with the replacement of Kruschev, in 1964 that things began to change. Lysenko was still allowed his institute, but modern views were tolerated, and Medvedev is representative of the revived school.

The detailed study of the twists and turns, the scientific fraud and the tragedy would again carry us too far from the original theme: Science and government. Rather, we should ask: How did this happen?

Medvedev addressed himself to this in his final chapter, and we can briefly summarize the reasons he lists: (1) A tendency upon decree of the government to classify particular sciences as either bourgeois or Marxist-socialist, (2) constant difficulties with agricultural production and the need for rapid improvement, (3) the role of the press inside Russia, which allowed some views to be supported and others suppressed, (4) the prolonged isolation of Soviet science from the west, and (5) the rigid centralization of scientific bureaucracy. On each of these, Medvedev elaborates, but even the bare listing here will indicate the complexity and interrelatedness of the forces which were in action.

Despite the thaw during which Medvedev was able to begin to air this review and which permitted the study of genetics to revive in an amazingly short time, relics of the older problems remain, and Medvedev himself was briefly confined to a mental hospital until protest by other Soviet scientists led to his release.[25]

8.6   Summary

To conclude this chapter, some general observations are in order. It must be expected that science will have increasing contact with government in the years ahead. Some of this will come through the need for expert advice, some because science is more and more dependent on the government for its finances. From all of this, there can emerge a science that is stronger and better able to contribute to immediate social problems, but there already exists and will increase the possibility for interference in and control of science by nonscientists. In the reverse direction, one can imagine a situation in which many major scientific/technical decisions are turned over to the scientists for their decision even though there may be important social implications. Neither of these courses seems attractive. Science and society need each other; neither can prosper alone, but neither can be allowed to dominate the other, and the Lysenko affair shows what can happen (admittedly under extreme conditions) when a science is made subservient in its content to ideology.

The independence of scientists, at least in the United States, is more likely to be threatened along the lines we have seen when describing the Oppenheimer case and HEW advisory panels, and scientists may well be victimized because of their nonscientific (mainly political) activities. While there are constitutional protections against some of these, it would seem that this should still be a matter for continuing concern on the part of the scientific community.

NOTES AND REFERENCES

1. Harvey M. Sapolsky, *Science*, **160** (1968), 280; Harvey Brooks, *The Government of Science* (Cambridge, Mass.: The M.I.T. Press, 1968), Chap. 4; Don K. Price, *Government and Science* (New York: New York University Press, 1954).

2. For example, Daniel S. Greenberg, *The Politics of Pure Science* (New York: New American Library, 1967); Don K. Price, *The Scientific Estate* (Cambridge, Mass.: Harvard University Press, 1965).

3. Galileo Galilei, *Discoveries and Opinions of Galileo*, tranl. and notes by Stillman Drake (New York: Doubleday and Company, 1957).

4. This generally held view of Galileo's adherence to the Copernican system has been critically challenged by W. Hartner, *Vistas in Astronomy*, **11** (1969), 31.

5. *Dialogue Concerning The Two Chief World Systems*, transl. by Stillman Drake (Berkeley, Calif.: University of California Press, 1953).

6. Loc. cit., ref. 3.

7. Giorgio de Santillana, *The Crime of Galileo* (Chicago: The University of Chicago Press, 1955); Arthur Koestler, *The Sleepwakers* (New York: Grosset and Dunlap, 1963); Ludovico Geymonat, *Galileo Galilei* (New York: McGraw-Hill Book Company, 1965).

8. For a good introduction to the growing literature on this subject, see Joseph Haberer, *Politics and the Community of Science* (New York: Van Nostrand Reinhold Company, 1969), esp. Chaps. 9, 10, and 11.

9. *In the Matter of J. Robert Oppenheimer*, Transcript of Hearing before Personnel Security Board (Washington, D.C.: U.S. Government Printing Office, 1954), 6.

10. Heinar Kipphardt, *In the Matter of J. Robert Oppenheimer* (New York: Hill and Wang, 1967).

11. The entire April 1955 issue of the *Bulletin of the Atomic Scientists* is given over to "Secrecy, Security, and Loyalty."

12. *New York Times*, April 17, 1969; *Science*, **164** (1969), 283 and **164** (1969), 406.

13. *Science*, **164** (1969), 532. *Bulletin of the Atomic Scientists*, **25** (1969), 2.

14. *Science*, **164** (1969), 813, 1499; **165** (1969), 269; **166** (1969), 357, 487; **167** (1970), 154.

15. Donald Fleming and Bernard Bailyn, *The Intellectual Migration* (Cambridge, Mass.: Harvard University Press, 1969).

16. A. V. Hill, *Nature*, **132** (1933), 952 and **133** (1934), 65; J. Stark, *Nature*, **133** (1934), 614; also J. B. S. Haldane, *Nature*, **133** (1934), 65.

17. J. Stark, *Nature*, **141** (1938), 770.

18. Haberer, loc cit., ref. 8.

19. For a review of the earlier known facts of this case, see Martin Gardner, *Fads and Fallacies in the Name of Science* (New York: Dover Publications, Inc., 1957) and David Joravsky, *Scientific American*, **207** (November 1962), 41. See references 20–24 for details as well as for information that has more recently become available.

20. *The Bulletin of the Atomic Scientists*, **4** (1948), 66, 70, 74; and **5** (1949), 130.

21. Zhores A. Medvedev, *The Rise and Fall of T. D. Lysenko*, transl. by I. Michael Lerner, with editorial assistance of Lucy G. Lawrence (New York: Columbia University Press, 1969); see also reviews of this book by Oscar Hechter, *Bulletin of the Atomic Scientists*, **26** (1970), 54, and T. Dobzhansky, *Science*, **164** (1969), 1507.

22. Theodosius Dobzhansky, *Bulletin of the Atomic Scientists*, **8**, (1952), 40.

23. Medvedev, loc. cit., 19.

24. See also David Joravsky, *The Lysenko Affair* (Cambridge, Mass.: Harvard University Press, 1970), and a review of this book by Frederick C. Barghoorn in *Science*, **172** (1971), 929.

25. *Nature*, **227** (1970), 1197, contains an extract from another book by Medvedev, to be published in 1971, entitled *The Medvedev Papers: The Plight of Soviet Science*.

values, priorities, and choice

chapter 9

Public attention in matters of scientific choice is usually focused on items that involve very large amounts of money. Over a period of several years, while the federal budget for research and development was expanding, the problem encountered might typically require a decision to be made between one major scientific project and another: A new particle accelerator or drilling a hole through the Earth's crust. More recently the emphasis has shifted so that a frequent question is now whether some major scientific project such as space exploration is more or less important than, for example, urban redevelopment or primary education or an expanded welfare program. Whatever the nature of the problem, the opinion of scientists can (or should) be one factor to be considered in reaching a decision. In an earlier chapter, there was some discussion of the role of scientists as advisers; here, we shall be concerned with the values that guide scientists in assigning priorities and thence in making choices. As has been done in several other discussions, the objective here is not that of arriving at definitive answers and values, but rather the airing of questions and the indicating of avenues which can be followed.

The day-to-day operation of science in the laboratory requires decisions. Some can be made relatively objectively on the basis of well established rules. So, for instance, there are standard statistical techniques that can be used to test which theory among those being considered gives predictions in closest agreement with the experimentally obtained numbers. On other occasions, however, the decision must be made on the basis of experience, which we sometimes prefer to call intuition. Such a situation might arise in the planning of further experiments: Which of several alternatives should be undertaken at all and then in what order. There is a real danger that this latter method of choice (relying on intuition or experience) can be quite biased in its operation, but an experienced scientist will often operate in this way, recognizing the risks but taking them in order to avoid a possible waste of time. Any experimental scientist becomes accustomed to making such decisions. In the planning of an experiment, in the discarding of bad data, in the emphasis to be given to the various parts of an experiment when describing them for publication—at all stages, decisions are being made.

Some of these decisions are made quite consciously, but some are probably made without the individual scientist's even realizing how deeply he has been indoctrinated into the prevailing values or theories. Even the choice of a scientific research topic is strongly influenced by a set of values now widely held by scientists and which Weinberg has very neatly summed up, in the form of four homilies:[1]

1. "Pure research is better than applied research."

2. "General is better than particular." (For example, a knowledge of the law of gravitation would be considered more important than recognition of a particular example, such as the falling of an apple.)

3. "Search is better than codification." (This is probably true for scientists to whom the search for new phenomena or laws is more interesting than the codification by which known laws are used to extend one's understanding of observed occurrences; the technologist, on the other hand, will often place a higher value on codification, which can be readily applied.)

4. "Paradigm breaking is better than spectroscopy" (by which is meant that confirmation of accepted laws and accumulation of more and more detail is not as useful in the extension of our knowledge as is the breaking of laws (paradigms), which can then presumably lead us to broader generalizations).[2]

The values which guide the scientific community change with time and reflect the values and felt needs of the general society. During the seventeenth century, when the scientific academies were being founded, practical uses for science were considered important, and in Chap. 7 we have seen how, during the Industrial Revolution, research was encouraged by the need for a better understanding of power and engines. While the research into the properties of heat was basic, its immediate daily application was clear. By about the middle of the nineteenth century, pure science began to be valued for its own sake, and the golden years of science around the turn of the century presented the public with an image of pure science at its most productive— discovery of many new chemical elements, the great period at the Cavendish Laboratory in Cambridge University, the discovery of the electron and of radioactivity, the remarkable 1905 papers by Einstein, and the experiments of Rutherford and the theory of Bohr, which together completely revised the ideas of atomic structure. University science curricula were greatly influenced and the prestige of pure science in the universities has persisted.

There is certainly no doubt that applied science and technology are still seriously handicapped through the present-day wide acceptance of the first of Weinberg's observations, "pure is better than applied." Until very recently there was a strong tendency for the best high school students to choose science rather than engineering programs in the university, and there has been a tendency for the better science graduates to prefer to look for academic positions (involving continued pure research) upon completion of their degrees rather than to go into industrial employment. Within physics, for example, the experimental and theoretical study of subatomic particles has attracted a disproportionate number of the better students, while atmospheric physics has been relatively neglected. Some of these fashions or biases are changing, but they have by no means vanished. Such values tend to be reinforced by the greater availability of research funds for some fields. We are now witnessing a reversal in funding priorities so that it is becoming easier to obtain federal research support in some areas of the applied sciences and correspondingly harder in some areas of basic science.[3]

This same value system can be seen to operate in another way. Students from developing countries who come to the United States for graduate studies very often are attracted to pure research. In many cases there will be no possibility for the continuation of this work when they go home with the result that many try to stay in the United States for as long as possible—a semipermanent brain drain. This has led to a marked division in the suggestions for coping with this problem, which is very real for a small country that faces losing many of its best students. One possibility is to establish some centers for basic research in the developing country so as to retain these better students when they have qualified. In this way they can contribute by teaching others who may not themselves wish or have the ability to go into research. Clearly this requires that at least minimal research facilities be provided preferably with travel allowances so that contact can be maintained with colleagues in other countries. On the other hand, there are many who feel that the support of this kind of pure research is a luxury for a developing country, and that all or most money set aside for science should go into establishing centers for applied research that will specialize in indigenous problems, ranging from tropical medicine to the use of local materials for low-cost buildings. While different countries are trying to cope with this problem in different ways, science is being judged on value scales that are set or maintained primarily by scientists in the already developed countries.

Another effect can be seen in engineering education. For many years electrical engineering programs were concerned with providing graduates who could fit into the electric power and consumer industries. Generator design and study of transmission lines were typically part of the curriculum. As radio developed, many departments began to offer a parallel program in what was often termed "light-current" engineering, and which would now be called electronics. As electronics became more sophisticated after World War II and especially in the 1960's, this field attracted more and better students. The field of high-speed computers is an offshoot of the great advances in this area. Concurrently, the old-fashioned heavy-current electrical engineering has appeared less and less glamorous and occupies a smaller part of the present programs. Although our economy is absolutely dependent upon the continued production of competent engineers in this field, and the country faces severe problems in the planning of adequate power supplies, it is becoming harder to "sell" this career to high school students about to enter college: Computers and microwaves seem far more glamorous. Thus do fashions and values in science and technology have far-reaching effects.

Whereas Weinberg's first criterion is concerned with the division *between* major divisions of scientific research, his other criteria neatly sum up the values that influence decisions *within* an area of science. When scientists are called upon to give advice in the ordering of priorities, it should not be surprising if some of these discipline-based values provide a source of bias. Why, though, should it be necessary to set priorities in science? Ideally one might expect or prefer science to take what directions it may, being guided only by the imagination and inventiveness of its practitioners. Those days are gone. By now almost entirely dependent on the federal government for its financial support, science must also depend on the decisions of various bodies for the allocation of whatever money is available.

At the highest level, it is Congress which votes the appropriations for those government agencies through which the funds will flow. The Administration, the Bureau of the Budget, the House, and the Senate all assist in shaping the final budget, and the amounts that come to science have been won in competition with the demands of other departments and agencies. The total amount available for scientific research is no longer enough to satisfy the more and more expensive tastes that have been cultivated, and choices must be made. More decisions must be made at lower levels. The National Institutes of Health, for example, have many panels. On these are included university scientists who review all applications for grants, and from these are selected those who will receive support. At another level, decisions of a slightly different kind are made. Within a university science department, for example, when there is a faculty vacancy, the question often arises: Should the department seek a theoretician in this speciality or an experimentalist in that one? Similar decisions confront those who direct the research in large government or industrial laboratories: How will the available funds be distributed among the various research groups?

Decisions affecting science are required at many levels, and scientists are usually involved in the administrative machinery. At lower levels where the amounts of money are not large, scientific advice and considerations are usually the controlling influence. At higher levels the decisions become much harder to make. But decisions *must* finally be made, no matter how hard it may seem to some scientists to have to make a choice. When the NSF is forced to decide between a new radio telescope and the proposal to drill through the Earth's crust, who shall decide which has greater scientific merit? Should the decision be left to a Congressional committee or to the Bureau of the Budget, with the possibility of rival engineering or industrial interests competing for the often attractive contracts? When decisions involve weighing one scientific project against another, it seems reasonable to expect science to provide the best advice. To a degree the National Science Board and the President's Science Advisory Committee can and do play such a role.

There has been little agreement so far, however, on what is really a prior question: How much can the country afford to spend annually on research and development? Paging through the "News and Comment" sections of *Science* for the past few years will reveal the ebb and flow of many suggestions, and we are no nearer to any set policy. At times, it has been suggested that the total R and D amount should be a fixed percentage of the federal budget. It has also been suggested that an attempt should be made to make allocations on a five-year basis, as is done in England, so that the universities and government laboratories are able to make intelligent long-range plans, and will also not be bound rigidly to annual budgets. An interesting proposal has come from Brooks,[4] who has suggested that it is only the total amount of money for academic research that should be considered competitively with the various other federal agencies; once a total is allocated for academic research, then presumably the division within that amount should be decided by the scientists themselves.

In the absence of any explicit agreement on the desirable level for total R and D support, it is not surprising that the pattern over the years should look somewhat random. Much basic research was supported for many years via the military research offices for which the Office of Naval Research provided the model. As the research funds for these military offices decreased in the late 1960's and their use was more narrowly defined, there was for several years no corresponding increase in the NSF budget. On the contrary, the NSF mandate has been broadened to include a greater coverage in the social and applied sciences. It is not the intention here to review the complex history of research funding in the United States, but rather to point to the difficulty in making sensible and long-range plans for scientific research when the rules of the game change during play. Nevertheless, major decisions have been made at various times where these decisions have major effects on scientific research policy but where nonscientific aspects have played major if not dominant roles.

The prime example is the Apollo program in which the manned spaceflight program was directed towards the landing of men on the moon and their return with samples from the lunar surface. By now this has been successfully accomplished as the culmination of a magnificent technological feat. The scientific results are still accumulating from the analysis of the lunar samples, and several further missions are in the advanced stages of preparation. The original decision, however, bears the clear imprint of politics. Mandelbaum, in an article in *Science*,[5] documented the background of this policy decision, while an earlier report by the Committee on Science in the Promotion of Human Welfare of the AAAS[6] had also discussed this part of the program. From both of these discussions, it is clear that military and political factors were more important than the scientific.

The total Apollo program is estimated to involve a cost of $25 billion; if this is taken as being spread over 12 years, then the annual cost is around four times as much as the total annual NSF budget. When the stakes are as large as this, it is not surprising that critical voices have been raised, both inside and outside of science. While scientists have been consulted at some stages of the program, many of the decisions are purely technical or strategic and will be made primarily to those ends.

Although the Apollo program has now reached the stage where almost all of the major decisions have been made, other programs of this kind or on this scale can be expected to arise, and the general problems still remain: When there is a prominent scientific component to a large program, how and at what stages should the opinions of scientists be incorporated? It must also be remembered that with political considerations so often being paramount, the choice is often not as simple as is sometimes supposed. Too often it is suggested that Grandiose Program A has insufficient scientific content, and therefore Grandiose Program B should instead be supported. This argument fails to recognize that funds denied to A have little or no bearing on the future of B. At the level of funding that we are considering here, projects rise and fall on their own, being the subjects of political moves that are beyond the control of the scientific community. As a specific example, if the space program were terminated today, it is quite unlikely that the NASA annual budget of close to $4 billion would suddenly be diverted to the NSF, the Office of Education, the Office of Economic Opportunity, or any other agency.

It is perhaps of more immediate practical interest to look at the smaller-scale projects, which may still involve millions of dollars but lie within the control of one or a few agencies. How is the balance made, for example, within the NSF between biochemistry and astronomy and geophysics and anthropology? Such decisions are not made at the Congressional level but will be influenced by the National Science Board and the NSF itself. Here again, Weinberg has provided a thoughtful proposal.

Weinberg considers "internal criteria" and "external criteria" for choice between fields.[7]

> *Internal criteria are generated within the scientific field itself and answer the question, How well is the science done? External criteria are generated outside the scientific field and answer the question, Why pursue this particular science? ... Two internal criteria can be easily identified: (1) Is the field ready for exploitation, and (2) are the scientists in the field really competent? ... Three external criteria can be recognized: technological merit, scientific merit and social merit.*

He then discusses these in detail, recognizing the difficulty in establishing criteria and in setting values such as social merit. While we may accept or reject all or some of Weinberg's arguments, they do represent a consistent and carefully explored attempt to establish some order and rationale from the present chaos and deserve far wider discussion than they have so far received.

Brooks has also set out criteria for choice:[8]

> *1. What is the promise of significant scientific results from the proposed project? The evaluation of such promise implicitly involves the past accomplishment of the investigator and the judgment of his competence and originality by peers, either nationally or locally. The term "significant" may refer either to scientific signific-ance or to potential applicability, but it implies some degree of fundamentality and generality.*
> *2. How novel is the work proposed? To what degree does it break new ground? To what extent does it exploit a new technique or unexplored research methodology? Does it provide a meaningful test of current theory and understanding in its field?*
> *3. To what extent are the probable results of the proposed work likely to influence other work either in the same field or in related or even distant fields?*
> *4. What is the probable educational value of the research, based on the quality and number of students or other trainees in relation to the cost of the project, the record of success of the investigator's students, and the general academic environment in which the work is to be done?*
> *5. What is the potential relevance of the work to possible future applications, especially to existing national goals? This question is of particular relevance in judging engineering research and applied research in health, agriculture, environmental pollution, or similar areas.*

Some of these overlap considerably with Weinberg's criteria, and this is perhaps not surprising since both Brooks and Weinberg have their pro-fessional roots in the physical sciences. When the merit of some scientific work (pure or applied) needs to be judged against a competing nonscientific project for the allocation of funds, then different criteria will surely be added. For instance, in support of various plans for bigger and better particle accelerators, some members of the nuclear physics community have become quite euphoric in describing their visions of the importance of these projects to society, but these outpourings have been treated with little respect.[9] Simply for eminent scientists to aver the importance of some proposed scheme is no longer any guarantee of its likely fruition. Until the federal research funds increase appreciably, one must expect major scientific research projects to be vigorously scrutinized by many who will not be applying the same yardsticks as will the scientists, and there will be increasing competition among the natural sciences, the social sciences, and the applied sciences. At the same time, more and more scientists and engineers are themselves coming to place more stress on the social criteria so that the revolution may well take place from within. Fashions change and so do values.

There is one further aspect to scientific choice which needs to be discussed before we close, and that relates to a basic feature of the granting of research funds. There are presently two main ways by which a scientist can gain support. In one he submits a research proposal to a funding agency (NASA, NIH, etc.) and this is judged for its scientific merit. The judgment may be made mainly by scientists working for the agency concerned or else be based on reviews by external scientists. The funding decisions must then be made within the agency, where the balance is set between different fields of research. This system is known as the project grant system. The other approach involves the award of a large single amount of money to a university, after a university-wide proposal has been reviewed. The university is then able to make internal decisions regarding the distribution of the funds, although some grants do contain some restrictions. The distinction between these two forms of funding is major and the implications are far-reaching.

The individual scientist is the person mainly responsible under the project grant. If he moves from one university to another, his grant will probably follow him. This system has been criticized as enhancing the mobility of scientists who may feel little allegiance to their university and are relatively free to move when it proves inhospitable or when an attractive offer appears. (Such mobility, though, is far from being confined to the sciences.) The merit of the system is that each project is judged for its scientific qualities and should in theory lead to the best scientific use of the funds. On the other hand, funds will generally follow the better scientists, and it will be harder for younger scientists to obtain support when funds are lean.

The argument in favor of institutional grants is also geopolitical. Large grants to geographically well-distributed universities can lead to a broader distribution of scientific talent, and can enable universities to attract good scientists. Furthermore, intelligent use of institutional funds can permit speculative but promising projects to be funded before they may be strong enough to compete for project funds.

At present, the overwhelmingly greater part of research funds is awarded through project grants, but many universities have campaigned for an expansion of the institutional system. The complete abandonment of the project grant system is strongly opposed by most scientists, who point to the difficulty that any institution will have when judging the competence of internal projects. An agency making project awards can draw on the best scientific talent available, but one might expect most universities to be subject to internal political pressures when the time comes to distribute internal funds. The debate continues.

Geopolitical factors also enter when decisions have to be made for the location of major national facilities. This includes major federal laboratories (such as those operated by the Bureau of Standards, or the National Institutes of Health), or laboratories connected with the space program (such as the Lunar Sample Institute in Houston). In addition to any scientific criteria, there must also be consideration for ease of access by air (for scientists who will need to visit or work), rail, road, and/or water (for the transport of exceptionally heavy loads); for the proximity to adequate electric power; and for the impact on a local economy.

As we survey the range of decisions that are continually being made, decisions which affect science either directly or indirectly we can see that more and more they are being made by nonscientists. Only at a relatively low level can decisions be made entirely on scientific merits; as one ascends the dollar scale, other increasingly important factors must be considered. Perhaps at some time, long past, science was able to decide its own directions, but now more scientists must come to recognize that many other legitimate demands are requesting support. One can expect that even the apparently completely internal criteria will come to reflect the changed views and pressures in the larger society.

NOTES AND REFERENCES

1. Alvin M. Weinberg, *American Scientist*, **58** (1970), 612.

2. In the sense used here, "spectroscopy" implies the painstaking accumulation of precision data in situations where the basic scientific laws are understood and where these tabulated data will be of practical use. Atoms and molecules can be identified by observation of the characteristic spectra they emit when heated, and comparison of their spectra with standard tables; hence the term "spectroscopy."

3. For a good summary of federal budget data relating to research, see Michael S. March, *Federal Budget Priorities for Research and Development* (Chicago: The University of Chicago Press, 1970); for a broader discussion of issues underlying the budget, see Harold Orlans, ed., *Science Policy and the University* (Washington, D.C.: The Brookings Institution, 1968). Regarding recent changes within the NSF, see *Science*, **170** (1970), 144.

4. Harvey Brooks, p. 62 in Orlans, op. cit. (ref. 3, above).

5. Leonard Mandelbaum, *Science*, **163** (1969), 649.

6. *American Scientist*, **53** (June 1965). See also John M. Logsdon, *The Decision to go to the Moon: Project Apollo and the National Interest* (Cambridge, Mass.: The M.I.T. Press, 1970).

7. Alvin M. Weinberg, *Reflections on Big Science* (Cambridge, Mass.: The M.I.T. Press, 1967), 71, 72.

8. Harvey Brooks, 71, op. cit.

9. Harold Orlans, *Science*, **155** (1967), 665.

responsibilities and obligations

chapter **10**

Science can not today be conducted in isolation from society. At one time it was possible for the dedicated individual to pursue his experiments at home or in his university laboratory oblivious to the real world outside. With few exceptions those days are gone. As has been discussed in the preceding chapters, the contacts between science and society are extensive and are expanding. Some scientists, either by inclination or by temperament, may still prefer to work quietly by themselves. However, from science as an institution and therefore from most scientists, society has now come to expect some participation, some concern, and some interest. For the scientists, the question is, therefore, whether their feelings of obligation and responsibility will come only grudgingly after prodding from the nonscientific society, or whether they will rather take the initiative and consequently probably play a larger role in the shaping of this mutual relationship. There are many facets to this relationship, reflecting different circumstances. Under some conditions, the scientist will appear as an expert whose objectivity is expected, but at other times the scientist will be participating in a very different capacity as an educated and informed citizen. It is essential that these two roles be clearly recognized and kept separate, and further that it be clearly spelled out which role a scientist is assuming at each time. The alternative is confusion, and out of this comes mistrust. Unfortunately, some scientists have on occasion blurred or failed to indicate the separation of these two roles, and so it might be considered a further obligation to draw attention to such practices when they occur.

In this final chapter, the discussion will be directed towards the area of the social responsibility of scientists. While particular emphasis will certainly shift from one topic to another with time, the underlying general concern will remain: Scientists must be sensitive to the possible or actual uses to which science is put, and they will be expected to participate in the debates which arise.

Increasingly, justified or not, scientists are being held responsible, as scientists, for the results of their research and the (mis-)uses to which they are put. Atomic scientists are blamed for the existence of nuclear weapons; chemists and microbiologists for the existence of chemical and biological weapons; engineers for the air pollution that largely comes from automobile engines and chemical plants. It is fashionable these days to lay the blame primarily upon the scientists and technologists who have made the original discoveries or the technical innovations, at the same time often ignoring the economic and social forces that have so quickly led to the exploitation of these discoveries, to the detriment of the environment, or to the increased horror of war. Are the scientists responsible? If so, to what extent?

Science is often put to uses that are far beyond the imagination of the scientist. Rutherford, one of the major figures in atomic physics research during the first quarter of this century, always held the view that no practical applications would come from the understanding of nuclear energy. Should Rutherford be blamed for Hiroshima and Nagasaki? The possibility of atomic weapons was even more remote at the time of the Curies' work on radium; should they be blamed, too? In such cases, it seems almost absurd to try to allocate responsibility so far in retrospect. The situation becomes rather different when the possibilities are indeed foreseeable. In many of the discussions today, it is suggested or implied that a scientist should not pursue research when the anticipated results may or will lend themselves to abuse (and the definition of misuse or abuse may itself be debatable). In some cases, the matter may be simply decided by the individual, who can exercise his freedom to seek employment in one place rather than another. After World War II, many physicists chose to return to university positions rather than continue with the development of further atomic weapons, while others chose to lend their skills to a variety of direct military applications. Since both decisions are quite lawful, attempts to persuade scientists to change their views fall within the political arena. It is worth noting, though, that in the mid-1950's a large group of German scientists did pledge themselves not to work on any nuclear weapons development.[1] There has been no corresponding gesture in the United States. On the other hand, there have been statements very recently from two major bodies that have dealt with chemical and biological warfare (CBW). In August, 1970 the International Congress on Microbiology, meeting in Mexico, adopted a statement[2] that urged "...all countries that have not signed or ratified the Geneva Protocol (on

CBW) should do so, and all installations...for offensive or defensive biological warfare purposes be converted to peaceful uses...." In September 1970, the Board of Directors of the American Chemical Society urged the United States Senate to ratify the Geneva Protocol.[3] These public stands are of considerable interest, in that they may well serve as precedents for other professional societies. The case of CBW differs from that of nuclear weapons. For CBW, there is the 1925 Geneva Protocol that prohibits the use of "asphyxiating, poisonous or other gases, and bacteriological methods" in warfare, while the uses of nuclear weapons are so far covered only partially by the Test Ban Treaty (1963) and the Non-Proliferation Treaty (1968). But these are in fact highly unusual statements on the part of professional bodies, and there is no statement, for example, promising not to design appliances which pose electric shock hazards to users. The contrast of domestic electric appliances with major weapons may seem ludicrous, and certainly the orders of magnitude are very different, but both involve the products of professional research and development. One may pose worldwide dangers if eventually used, yet the other may also pose dangers to many people, on a daily basis even if at a lower level.[4]

In some ways the cosmic subjects of nuclear and CBW weapons are easy to deal with, simply because of their very enormity. The apparently trivial or mundane matter of adequate electrical insulation is representative of a very large class of problems in which there are well-defined solutions, but where economic reasons (i.e., greater potential profits) dictate the cutting of corners. A few more examples will be of use. The technology of automobile tires is well advanced; tires that wear longer and are safer cost more than those deliberately designed for a low price. For a higher price, more safety features can be built into an automobile. Certain drugs are known to have major side effects, yet the complexities of the regulatory procedures permit a manufacturer to continue marketing during extended litigation. To whom are the scientists, chemists and engineers in these companies responsible, when they recognize hazards in their products? It is easy to claim that the first responsibility should be felt towards the users of defective or dangerous products, but for the individual scientist or engineer involved, his continued employment is probably at stake. We do not yet have procedures to protect such men, and we should also recognize that free disclosure of industrial information can seriously affect a company's competitive position. At present, allegiance is felt primarily to the employer and is demonstrated by silence.

Just how these commercial and governmental pressures can effectively inhibit the disclosure of information can be seen in cases that have been well publicized. The vigorous efforts to alert the public to the potential hazards of radioactive fallout came almost entirely from university scientists and doctors. Scientists who worked for the AEC (with the notable and very recent exceptions of Gofman and Tamplin) were either silent or tended to minimize the concerns. In the debates on civil defence, and later the ABM, it has again fallen to university scientists and concerned citizens to raise objections. As environmentalists have become increasingly aware of the many unsuspected side effects of herbicides and pesticides, scientists in industry have either been silent or else have defended their products. In a few cases, industrial scientists who have been publicly critical of a product have been quietly penalized. University scientists have been protected by their academic tenure and the greater tradition of freedom to speak out. There are instances where scientists in one government agency have been able to take public stands that are critical of the safety of the program of some other agency, but these are the exceptions. Overall, academic scientists have greater freedom, but they often do not have access to information that is available only to those in industry or government from whom critical comments are unlikely.[5] (In issues of the type described here, the scientists alone would have been unable to play the role they did without the support of many commentators in radio and television, and editors and reporters in the press. It is not intended to belittle the contribution these have made but rather to contrast the actions of independent scientists with those in government and industrial laboratories.)

Questions of professional responsibility and the misuse of knowledge do not confine themselves to the natural sciences and technology. In the social sciences, research has yielded an improved understanding of human behavior, both individually and in social groups of all sizes. From this knowledge have come many applications with wide use. Companies use public relations consultants and direct their advertising with care; consumer demand is manipulated; political parties make use of similar methods in major election campaigns; governments use propaganda both internally and externally. In these and thousands of other applications, to whom do the participating psychologists and social scientists have prime responsibility?

As has been discussed in Chap. 5, professional societies have tended to avoid social involvement. Some are now providing sessions at their meetings at which social problems are discussed, but the societies are still not taking majority positions. The matter of the Geneva Protocol, referred to earlier in the present chapter, is quite unusual, but it is perhaps not unfair to suggest that even those resolutions might not have emerged had there not already been almost universal condemnation of the use of chemical and biological agents in warfare. One might then go further and wonder what use such resolutions serve in a situation where there is such general agreement?

While the strictly professional bodies continue their main objective, the promotion of their respective sciences, a new class of organization has sprung up reflecting the need felt by many scientists for greater social involvement. The Federation of American Scientists (FAS) and the Society for the Social Responsibility of Scientists have relatively small memberships, amounting to only a few thousand, but both print brief news sheets for members, and the FAS in particular has taken a vigorous stand on many political issues. Even newer are groups that are more radical politically, carrying such names as Science for the People. It is too soon to know what impact they will have, but their effects may depend upon their tactics. Some of them persistently interrupted the AAAS meeting in Chicago, in December 1970—a tactic unlikely to win friends and influence people today.

Adopting a very different position from both the professional bodies and groups such as the FAS, there is yet a third class of organization, in which scientists are concerned only with providing accurate information in a form understandable to a broader public. This movement started in 1958 when there was much public concern and discussion over the matter of the biological effects of radioactive fallout from weapons tests. To a very large extent, the public had little or no accurate information easily available—on the nature of radiation and its biological effects. In St Louis the Committee for Nuclear Information was formed, including both scientists and concerned laymen. With time the interests of this group have broadened to include all environmental problems, and the name has changed to the Committee for Environmental Information. A monthly magazine is published (*Environment*) with detailed and documented reviews so that the interested nonexpert may inform himself. In addition, the committee provides speakers for local groups and has also testified before Congress and a variety of hearings on environmental problems. The basic motivation stems from the belief that rational decisions cannot be made by an uninformed public. Furthermore, the public needs information which is presented by an unbiased source, which will not seek to endorse policies but rather point to the potential or actual risks and benefits that can be anticipated from various courses of action. Similar information groups now exist in many parts of the country and are affiliated to the Scientists Institute for Public Information, which maintains offices in New York.

For a long time, the scientific community has been involved in bringing science to the public, but this has usually been confined to explaining new discoveries (for instance, how a laser works) or in lecture-demonstrations involving scientific phenomena (often at the gee-whiz! level). What is quite different with the new information groups is their attention to explaining the scientific content and implications behind major issues that confront the public. This still leaves plenty of room for popularized science, on the one hand, and political support and advocacy, on the other.

This involvement in reaching a broader public has other aspects too. From the late 1950's, many active scientists have participated in numerous projects to improve the science courses offered at elementary and high schools.

Greatly improved texts and laboratory equipment have emerged often from collaborations between scientists and teachers. This is not to say that all of the new curricula are equally successful nor are the subjects such as physics and mathematics necessarily made easier, but modern and accurate materials are now more widely available, including a greatly extended range of audio-visual aids. Informing the public requires many different approaches, and the educational aspects are extremely important since attitudes towards science are often set before or during high school. If school science curricula have the effect of alienating many students from science, it will be difficult to persuade those students in their adult years to take any serious interest in public issues that involve science and technology, no matter how important those issues may be.

In discussing the responsibilities of scientists, attention has focused on the social obligations, but for completeness it should at least be mentioned that there are also responsibilities internal to each science, more properly falling within the area of professional ethics. Every working scientist is involved in making decisions that reflect this problem: In publications, there must be adequate acknowledgement to others, whether for specific ideas or assistance in various ways. Ethical problems arise very quickly when human subjects are used for research in medicine and the social sciences. Physical or psychological damage can result despite careful planning, and generally it is expected that the subjects of the proposed research will give their informed consent before participating. Confidential information is often obtained under the promise of anonymity. All of this places a heavy responsibility on the scientist, and careless or deliberate departures from generally accepted standards can not only hurt the subjects but also jeopardize public confidence in further research. The United States Public Health Service, which finances much of the research involving human subjects, has therefore drawn up guide lines for the conduct of this research. These require review of plans and procedures by a panel to ensure that[6]

(a) *the rights and welfare of the individuals involved are adequately protected,*
(b) *the methods used to obtain informed consent are adequate and appropriate, and*
(c) *the risks to the individual are outweighed by the potential benefit to him or by the importance of the knowledge to be gained.*

When problems occur, alleged breaches can sometimes be handled by the appropriate professional bodies (such as the American Medical Association), but there have been instances when the repercussions have extended further. It is accepted that in some situations involving personal interviews full disclosure of the purposes of the survey may prejudice the answers to questions. The good faith of the interviewer is then of prime concern, and prior review of aims and proposed methods by a screening panel would seem essential.

We have come to feel that the essence of intellectual and scientific freedom is the ability to investigate subjects of one's own choice, but this is becoming increasingly more difficult in the social and behavioral sciences. There may well develop an inhibition against research in some areas, for instance into behavior considered socially deviant, because of the difficulty in ensuring adequate safeguards against disclosure and identification. Some researchers object strongly to limitations being placed on the conduct of their research. Should they be free of supervision in their work? Are the potential results of such studies of sufficient importance to counterbalance intrusions into the privacy of individuals and the possible harm in the event of disclosure? The PHS guidelines are designed to deal with these problems in such a way that research is neither unchecked nor unduly restricted. But we at once come across a problem of implementation. If one considers issues of privacy and the balance of risks and benefits to be an ethical question, broader than just the professional discipline, should the reviewing panels be restricted to professionals close to the field concerned, or should there be provision for the inclusion of other educated persons who have ethical concerns?

There is another aspect of responsibility that confronts the individual scientist in his own work. Should he continue with some research when he can see that it can be used in ways of which he disapproves? In discussions amongst nonscientists, there seems to be an implicit assumption that refusal of a scientist to divulge the results of his research will mean that these results will not become known. The very nature of science denies this possibility. Refusal to publish can delay the introduction of some new techniques or process, but historic instances of repeated discoveries show that scientific advances cannot be so easily avoided. Although scientists who are concerned with possible misuse of their discoveries may prefer to work on different problems, they are clearly the ones in the best position to draw public attention to these possible misuses and to help to ensure that they will not occur.

The use of nuclear explosives for military purposes is probably what most people would cite if asked for an example of the misuse of a scientific discovery. In some ways, however, this particular case is not comparable to the problems we now face. Nuclear energy was developed from its laboratory to practical usage under the stress of an unusual war. The forces that then drove scientists to develop this weapon are now so far in the past as to be almost incomprehensible to the new generation of students. Today some very difficult problems can be foreseen, not within a context of their possible use in an unpopular war but rather for the effects they can have on the whole society during a time of peace. For example, depending on the outcome of further research, it may prove possible to select the sex of a baby before conception. Other possibilities of genetic manipulation may produce children with selected characteristics or abort those with potentially dangerous traits.

Perhaps these possibilities will remain within the pages of science fiction, but if they do become realities, who shall make the decisions? Should they be left to the preferences of individual parents, or should a broader segment of society set guide lines? Some scientists consider these possibilities so remote as not to be worth worrying over; others feel that the pace of research has been so rapid that it is worthwhile to start considering possibilities and alternatives. As with so many other questions, should this be left entirely to the experts?[7]

As has been mentioned earlier, there has been criticism of scientists whose research is supported by a military agency, even though the research is not secret, and the results may be openly published. Should the scientist do none of this work at all, even though there may be many important purely scientific applications? For instance, some mathematical statistical analysis may find uses in many branches of science but can also be used for the analysis of bombing patterns. Is this sufficient reason for all mathematicians to refuse to work on these problems? In any event it is highly likely that there will be mathematicians who do indeed agree with the military strategies and who therefore are quite willing to undertake the corresponding research. Should other scientists refuse to use these mathematical techniques simply because they were developed for military purposes? These questions are posed here, neither rhetorically nor for obtaining easy answers but rather to point to issues that are now being raised and to suggest that there are neither obvious nor universal answers.

How far should the idea of responsibility be carried? Siekevitz[8] has recently raised the question of possibly introducing professional oaths somewhat along the lines of the medical profession. This suggestion has been made before, but the feelings of social responsibility are far stronger now and more generally recognized. However, with or without an explicit oath, what should be the attitude of professional science to one of its members who departs from professional standards? Should there be more restrictive membership qualifications for professional bodies? Siekevitz goes further and suggests that there should be a

> ... *standard of responsibility. Such a standard would not allow us, for example, to call physicians those who would experiment on humans; to call microbiologists those who would develop lethal viruses; to call chemists those who would develop chemical weapons; to call physicists those who would build bigger and better bombs.*

Siekevitz suggests that

> ... *scientists who consistently use their skills in the service of killing men ... should not be asked to meetings, and should not be allowed to publish their results. They could be free to do their work, but I think we have a right and a duty not to acknowledge them as fellow members of the scientific community.*

As might be expected, other scientists have different views on the matter of professional responsibility. Chain,[9] for example, would appear to hold a position similar to that stated by Da Vinci and quoted earlier in which the protection of liberty is considered a prime obligation. In Chain's view,

Throughout the ages, wars were won by superior weapons. In modern war technology based on science plays a bigger factor than ever before in the history of mankind. Capable scientists are, therefore, the most precious asset which a nation possesses to give it superiority over its enemies, and victory or defeat is in their hands.

Consistent with this, Chain states,

No society can permit its scientists to back out in case of an emergency when its very life is threatened, and no scientist is morally justified when called upon in such an emergency to play his part to deny his services to the nation of which he forms a part, and which defends him and his family.

This naturally leads to the question of secret reseach within the universities, for military purposes. Chain considers there should not be such work on the campus in times of peace, and his reason for this is "...simply because in university surroundings there is not sufficient security to keep important discoveries secret." In other comments, Chain points to the obligation of scientists to help inform the public as to possible uses of science and technology, but he places the prime responsibility on society itself for the uses it makes of these discoveries.

The contrast between the views expressed by Siekevitz and Chain illustrates very well the polarities of approach to the role of science and scientists. Many scientists remember World War II and the desperate feeling that united the opposition to Germany and also led scientists deliberately to try to turn a discovery (nuclear energy) to military ends. The fear of many was that Nazi rule would be extended through conquest, possibly with the aid of nuclear weapons. A similar fear or at least a mistrust of the major communist countries appears to motivate many today. Chain's views on the need for defense seem similar to those propounded by some scientists in this country who have strongly supported nuclear weapons, ABM, and civil defense programs. On the other hand, Siekevitz expresses views probably held by a growing number of younger scientists in the United States, who too young to know World War II, fear not conquest but nuclear annihilation.

It is not at all clear that these diametrically opposed views can be reconciled. The main lesson to be learned at this juncture is that, as usual, there are no absolutes in scientific morality nor even a code of acceptable practices. As a result the scientific community is quite different from the legal and medical professions in having no code of accepted standards against which members can be judged. Perhaps one might consider that the scientists should make a start toward formulating some such code and at least start discussing the issue. As things now stand, many scientists are probably quite unaware of even the existence of many of the issues that have been raised in this chapter.

In discussing the social responsibility of scientists, we have dealt with issues that can arise in the normal course of scientific work: The use or misuse of some new discovery; how to inform the public; the reconciliation of research progress with the rights of human subjects of research; the obligation of a scientist to help with the defense of his country. A broader view of responsibility can be taken if one starts from a different position, and some of the particular issues listed above will then emerge as parts of this broader view. One can take as a starting point the opinion expressed by some that society now faces many severe problems, some generated by the applications of science. In this view what is needed is a major effort on the part of science now to avert worse disasters in the years ahead. Platt[10] has explored such a view and has attempted to assess the magnitude of various problems and prospective crises. These range from total annihilation in nuclear war down through major urban difficulties (transportation, crime, neighborhood problems) to medical problems. For each he has tried to make a quantitative estimate of the intensity of the issue (which he defines as the number of people affected times the degree of effect), and the time within which substantial success is needed to obviate crisis. His conclusion is that a major mobilization of scientists may be needed. If this action seems improbable, it is well to remember that such a mobilization requires that the scientific community come to feel the importance of the issues and the need for action. Essentially this is what happened in World War II when a major part of the scientific and technical expertise was concentrated on the war effort to produce nuclear weapons, radar, and other weapons and aids. Perhaps the growing demand for scientific relevance will turn in the direction that Platt has explored.

This brings us to the final point to be discussed. C. P. Snow[11] and many others have commented on the general theme of the existence of two cultures, although the actual number of cultures has been disputed with suggested numbers running from zero to several. The main point, however, is well taken. In a society that is heavily dependent on science and its applications, surprisingly few of the public have any understanding of the conduct of science or even a few of its major ideas. Snow deplored the existence of this state of affairs in which a man could be considered cultured if he was conversant with the classics and the arts, but remained ignorant of science. Today with major political decisions involving technological issues, it is clearly in everyone's best interests that there be a more widespread understanding of science and its applications. The responsibility to help achieve this must be recognized and assumed by more of us (both scientists and nonscientists) if we are to continue to benefit from the progress and promise of science. What is needed is a much broader participation by scientists in helping the public understand how science operates and what it can (and cannot) do. This participation can take many routes, but there will be no progress until the scientists themselves come to feel the importance and urgency of this task and can then persuade others that their efforts will be worthwhile. It is in the hope of at least stimulating some of these discussions that this book has been written.

NOTES AND REFERENCES

1. *Bulletin of the Atomic Scientists*, **13** (1957), 228.

2. *New Scientist* (September 10, 1970), 518.

3. *New York Times* (September 16, 1970).

4. *National Commission on Product Safety; Final Report* (Washington, D.C.: U.S. Government Printing Office, 1970).

5. See *Science*, **171** (1971), 549 for a report of a meeting organized by Ralph Nader at which many of these topics were discussed together with possible courses of action.

6. *Protection of the Individual as a Research Subject* (U.S. Department of Health, Education, and Welfare, Public Health Service; May 1969). For a detailed survey of this whole area, see also *Daedelus* (Spring, 1969), where the entire issue is devoted to essays on "Ethical Considerations of Experimentation with Human Subjects."

7. See editorial in *Nature*, **231** (1971), 69, and article by Robert G. Edwards and David J. Sharpe, "Social Values and Research in Human Embryology," *Nature*, **231** (1971), 87.

8. Philip Siekevitz, *Nature*, **227** (1970), 1301.

9. Ernst Chain, *New Scientist* (October 22, 1970), 166. This article was first published in *New Scientist*, the weekly international review of science and technology, 128 Long Acre, W.C.2, London, England.

10. John Platt, *Science*, **166** (1969), 1115.

11. *The Two Cultures and a Second Look* (New York: Mentor Books, New American Library, 1963).

appendix

1. Editorials and correspondence from *Physics Today*, pertaining to the Schwartz proposed amendment to the constitution of the American Physical Society. (Reprinted with permission of the editor of *Physics Today* and the authors.)
    Editorials: December 1967, August 1968
    Letters: January, February, April 1968

Physicists and Public Policy

A while ago the editors of *Physics Today* rejected a letter for publication that was concerned with the right or wrong of the Vietnam war. We gave as our reason that the American Institute of Physics and its publication *Physics Today* are by charter and intent devoted to physics as physics and physicists as physicists. The letter did not appear to have any special relation to either of them.

The rejection caused a flurry of correspondence, much of it with officers of the American Physical Society, largest of the seven member societies that are federated in AIP. The officers were urged to use their influence with AIP to change the decision and also to make APS meetings and publications available for discussion of public issues like the Vietnam war.

The matter has now come to a point of decision as we reported in our November issue on page 81 and as we report in this issue on page 69. Those who feel physicists should be able to get involved in public issues through their societies have proposed an amendment to the APS constitution to increase such involvement. The entire membership will vote on the matter by mail after discussion in *Physics Today* and at the January-February APS meeting in Chicago.

More precisely the amendment would allow 1 percent or more of the APS membership to initiate a vote by all members of the society on "any matter of concern to the society." In view of the purpose stated in the APS constitution ("advancement and diffusion of the knowledge of physics"), just what matters are "of concern to the society" is still somewhat open. Presumably the society council would have to decide. But the drafters of the amendment are clearly interested in raising questions of public policy not necessarily closely related to physics.

Arguments in favor of the proposed amendment deserve appreciation. Professionals in prewar Germany, for example, are frequently damned for not using their influence to stop the trend toward gas chambers. Physics is already deeply involved in the lives of nations, and its discoveries have broad technological and social implications in both peace and war. If physicists as a group can clarify a problem or suggest a solution not clear to others, they should have the means to do so. Recognition of the need for a means to meet such problems was one of several reasons that led AIP to form its Committee on Physics and Society (COMPAS) (*Physics Today*, August, page 53). COMPAS is designed to be an interface between physics and society. All of these arguments urge discussion of public issues by large numbers of physicists at meetings and in publications.

But contrary arguments also exist. One is the risk that meetings and publications now accomplishing their purposes fairly well may be badly diluted or disrupted by injection of issues not now on the agenda.

Another is the danger of ineptitude and arrogance. The essential nature of physics is that it deals with simplified models of simple systems (point masses, hydrogen atoms, the compound nucleus) and seeks exact solutions where such solutions are available. People trained to deal with such models may not have any special skill in dealing with more complex systems like the human mind and international politics. They might offer a lot of bad advice. Public assertion that physicists have special competence might be an arrogance that would bring discredit to a community in place of the respect it now enjoys.

A third danger is misinterpretation. If a vote of physicists produces a resolution based on a majority opinion, that opinion is likely to be quickly identified by the public as the opinion of the entire profession. The minority view will be inadequately represented; the members of the minority will be upset.

A corollary to these last two arguments is that legislative matters relating to physics may come to be decided on political rather than scientific grounds.

Vast projects and federal support of physics may gain or lose because of physicists' views of administration policy.

The injection of political issues into scientific meetings may ruin a situation that scientists and statesmen have labored long to produce. Today a scientist of any political persuasion from any country can get approval to come to scientific meetings in the United States. Contacts between eastern and western physicists have gradually increased for several years while both eastern and western nations have relaxed the regulations that, in times past, frequently stood in the way. These welcome contacts may cease if what were once pure physics meetings become political.

Complications, too, may threaten any move to take up public issues not specifically related to physics. For example, can an organization chartered for the advancement of physics legally make pronouncements on other matters? How far out of its declared path can a society stray without jeopardizing its tax-exempt status? If the nature of any society is to change, should a new society grow up to fill the role that the society used to fill before the change?

If deeper involvement is appropriate, several methods deserve consideration. Already available for opinions are other organizations and media, some of them (Federation of American Scientists, *Bulletin of the Atomic Scientists*) especially devoted to scientists' views. In response to the present move, APS and other physicists' societies might open their meetings and publications fully to debates and pronouncements on all subjects. Alternatively they might allow their governing bodies (in APS, the council) to determine what subjects are acceptable. Another possibility would be that a separate organization or a special section of a society be formed to deal with matters far removed from the central concerns of physics.

Perhaps physicists need their own channel for expression on public issues. Perhaps, too, they need to observe the differences between their roles as physicists and their roles as parents, citizens and human beings. Whatever the need, those who seek to fill it should look carefully at what they stand to lose while they are seeking routes to what they hope to gain.

The Editors

Should APS Discuss Public Issues?

FOR THE SCHWARTZ AMENDMENT

As the author of the constitutional amendment now before members of the American Physical Society I would like to present arguments in favor of its adoption. There are two questions to be considered. The larger one is, Should the American Physical Society involve itself in public issues?; and the specific one is, Why is this constitutional amendment needed? Let me start by answering the second, more technical, question.

One individual physicist may talk to another about any subject at all, but if he wishes to address the entire membership of his professional organization he must have the approval of those officers of APS and the American Institute of Physics who control the publication facilities. While I agree that some controls are needed, recent experience has shown me that the present manner in which these decisions are made is seriously out of balance. I believe that there operates today a censorship completely alien to the principles of free discourse upon which a scientific community is built. The correctness of this opinion is most clearly demonstrated by the manner in which the debate on this amendment has been handled. The editors of the *Bulletin* of APS and of *Physics Today* have rejected publication of both a summary statement and a thorough expository article, by means of which I had hoped to explain to the society membership at the outset of the debate just what had motivated 248 members to sign the original petition. Instead, and against repeated objections, they have chosen to present this whole debate in their own terms, as if they could play the role of an impartial mediator, when in fact they represent the chief target of my complaints. By the time this letter appears in print—at least two months after the first announcement of the proposed amendment—I fear the issues may have become badly confused.

The change we hope to achieve should lead to a more open-minded attitude on the part of the society towards new situations now and in the future. In the opening sentence of the proposed amendment, "The members may express their opinion," etc., the emphasis is on "members." The basic idea is that the members retain for themselves the right to decide which issues they wish to consider and which they choose to ignore. Specifically, upon petition by 1 percent of the membership any question, in the form of a proposed resolution, would be placed before the society for formal consideration and voting in a mail ballot. This critical measure, 1 percent, should make it not too easy for any extreme faction to coerce the majority, but not too hard for a respectable minority to get its views presented.

Further discussion of the procedural details of the proposed amendment is, I understand, the subject of other items in *Physics Today* and so I shall not dwell on these here. However, one crucial point deserves comment: the interpretation of the phrase "on any matter of concern to the society," which defines the scope of resolutions that members may vote upon. The editorial in the December *Physics Today* says that presumably the APS council will decide how to read this. While I agree that the council might concern itself with this question, I point out that the whole intent of this amendment is to create motive power for the members outside of the council. Thus I claim the view should be that any matter meeting the formal requirements (1 percent support) was *ipso facto* of concern to the society.

Now I turn to the major question of society policy: the appropriateness of discussing public issues. Certainly one of the easiest ways to destroy the integrity of the society would be to turn it into a debating club open to every political issue of the day; and the proposed amendment is carefully designed to protect against such excesses. At the other extreme we must recognize the absurdity of complete political innocence. Such statements as, "We are concerned only with physics as physics," are simply nonsense. There exists a whole range of issues where the technical activity of physicists gets tied up with political decision making. Our individual requests for government funds and the scientific appraisal of others' proposals are the most obvious examples. Each reader, and each letter writer, will doubtless have his own list of priorities in this regard. The choice in these cases of whether to take a position—as a professional group—and when to stand aloof should always be an open question, to be decided by the members as a whole once some threshold of community concern has been passed. At present it too often happens that the "public opinion of physicists" emerges from sources quite remote from the actual majority of our colleagues. (For this we have only our own lassitude to blame.)

There is one other situation when, I believe, my professional society should concern itself with a public issue: when there exists an external crisis of such magnitude that we fear a general catastrophe of a political, military or cultural nature. In my view the Vietnam War in all its ramifications does now pose such a crisis; and I would like to see the Physical Society face up to this issue, not because we have any unique competence in this matter, but because we share an equal concern and responsibility along with all other segments of American social structure.

In closing I return to the immediate question of the proposed constitutional amendment and remark that it refers to no particular issue or class of issues. It simply seeks to establish the means whereby the members can take it upon themselves to consider when some issue may be pertinent to their professional future. That is to say we are individually and cooperatively willing to be responsive to external realities, while retaining concern for our internal integrity as scientists. Such a commitment is neither easy nor guaranteed safe from criticism, but I believe it is a responsibility we should assume. If not, then we shall continue to be judged according to the dictum, "silence implies consent."

Charles Schwartz
University of California, Berkeley

Editors' Note: We do not wish silence to imply consent to all of the statements in the preceding letter. The following is the closing paragraph in a letter that we sent to Charles Schwartz on 11 October:

Consequently I return to you the article that you recently submitted and suggest that you make it into a letter to be part of the correspondence we will publish. In the letter we would like to have your own point of view as expressed in most of your article. We do not need the text of the resolution because it will appear elsewhere; we do not want the points of view of other people that you quote because they will have their own opportunity of expression; we do not want your "Exhibit A" and "Exhibit B." We do not need exhibit A because in the foreseeable future we do not plan to discuss political issues, and we do not want exhibit B because it is available in the APS con- stitution and it has been briefly summarized in our November story. I look forward to receiving a letter from you.

In a subsequent telephone call our chief editor told Schwartz that we would remove the strictures placed on him in this paragraph and allow him to publish anything he wanted to say in 3000 words including the original letter that was rejected. The editor reiterated that he thought Schwartz would dilute and weaken his argument by including other unnecessary material but allowed him that privilege if he wanted to use it.

The basic purpose of the letter was to invite submission of what Schwartz refers to as "a thorough expository article," in a form suitable for this January "Letters" department. The overall delay would have been about four months from date of submission, and such treatment is not approp- riately labeled by the words "rejected publication."

We of *Physics Today* and AIP do not feel that we "present this whole debate in (our) own terms." Perhaps the letters in the present column and our December editorial are the appropriate evidence on which to judge.

Let us call to your attention that the quoted, "We are concerned only with physics as physics," is a misquote to the extent that the word "only" has been added and appears to give an emphasis that was not intended in the original editorial.

Finally we do not wish to consent by silence to the term "censorship." Because of limits to space and staff, we must make choices. The terms of reference for *Physics Today* and AIP are stated monthly on page 6 of this magazine: "advancement and diffusion of the knowledge of physics."

As we make our judgments in following this purpose, we are sure we make errors. We do not feel, however, that censorship is properly classed among them.

The Editors

RESPONSIBILITY TO SOCIETY

It is generally agreed that scientists have two traditional duties: first, the duty of seeking the truth; second, the duty to communicate to all who need it the knowledge gained in their search. Because of our burgeoning tech- nology, we have reached the point where many key political judgments must be based on technical knowledge and scientific judgment. These are judgments that should not be left to the politicians who on the whole are seriously lacking in scientific background. Scientists who help contribute toward political judgments in this computer age are performing a valuable public service and should be encouraged in this by their societies and publications. I am not advocating discussion of scientifically related political issues in *The Physical Review:* however, *Physics Today* appears to be suited for this.

Another duty we scientists have is to protect the public from scientific hoaxes, whether small or large. This can best be illustrated by going into a specific example. It is purported that the President's scientific advisers and the Secretary of Defense have recommended against deployment of a $5 billion ABM system. There is a case to be made on scientific grounds that the system will not work as advertised. Should not the pages of *Physics Today* be open to knowledgeable physicists who are deeply concerned about such issues? Certainly there is a lot of physics in how high-altitude nuclear explosions can be used to blanket out long-range radar. This one technical point that has so far not been openly discussed could possibly be the death blow to the proposed thin ABM system. An open discussion among independent scientists as well as the "establishment scientists" will better prepare us for exposing this hoax, if it is a hoax.

Not only should *Physics Today* help us fulfill our responsibilities to society, but the APS should be providing for one or two sessions or symposia per meeting on science and society. The AAAS and its publication *Science* are way ahead of us. This year's annual meeting of the AAAS will have two sessions on the ABM issue among others of political interest. *Science* has had lively letters to the editor on chemical and biological warfare and the role of the scientist in the Vietnam conflict. Now that science has so swiftly spread to almost every corner of our society, I would recommend a similar editorial policy for *Physics Today*.

The original APS constitution was fine for the age in which it was written. But since then the atomic age, the space age and the computer age have exploded upon us. Do we not have a duty to our benefactors, the taxpayers, to at least keep up with the times?

Jay Orear
Cornell University
Chairman, Federation of American Scientists

JOURNALS IN OTHER HANDS

I am interested that a vote or poll is in the process of being taken to determine whether some of the AIP journals, most particularly *Physics Today*, should be opened up to debate on issues such as Vietnam and the like. It seems to me that it would be a great mistake for the American Institute of Physics, or the member societies, to be involved in such controversial matters unless they are tied very immediately to the profession of physics itself. To use our journals for very general discussion and debate means that they will lose their essentially professional character. Still further, the journals will fall into the hands of politically or socially oriented editors who will inevitably use them to support their own special viewpoints on matters far outside of the field of physics.

Every physicist is a citizen and has countless avenues outside his professional journals for expressing social and political views not immediately related to his profession. If his views have special merit, he will reach a far wider audience and hence be much more effective by using the broader media for presenting and supporting his opinions.

I wish to emphasize that I see no reason why *Physics Today* could not be used for comments on social or economic issues immediately related to our profession. On the other hand, such activities should be exceedingly well conceived if they are to have special meaning. By and large, one should strive toward something like the level approached in *Nature*. It is not easy to find editors who are capable of initiating and sustaining good analysis of this type. It is better to shun such areas than to enter into them badly.

Fredrick Seitz
National Academy of Sciences

SEMANTICS AND SUBSTANCE

The proposed amendment to the APS constitution should be defeated not necessarily because of its merit but because its wording is very vague and ambiguous. Presumably the crux of the present discussion is to define what is and what is not "a matter of concern to the society," and this is left undefined in the amendment. Passing it in its present form would open the door to discussion within APS of any issue under the sun, a discussion that would be hopelessly divided between the merits of the particular issue and the appropriateness of the issue as a matter of concern to APS.

Speaking of the substance of the amendment, I would be in favour of a change that would allow a discussion within APS of issues directly pertaining to the work of its members as scientists. Such issues would include topics like scientific priorities, scientific aspects of foreign aid, the impact and the connection of science with governmental agencies such as the Department of Commerce, Department of State, Department of Housing, etc., the allocation of national resources for scientific research, the social structure of scientists themselves, and many others. I feel that these are issues that at least partially match the special competence we are supposed to have; therefore their consideration would be a significant contribution to ourselves and to society as a whole.

I would be very much opposed to allowing APS to discuss general public issues. Each of us, as an individual, has innumerable outlets for such activity, including some organizations especially geared to scientists. Dragging APS into this would be quite superfluous. In addition, I am also very sensitive to any trend that might result in public statements by APS on issues in which we have no special competence. Such statements would necessarily carry in the public mind a false tinge of importance and authenticity just because they are utterances by scientists who, according to the contemporary myth, have at least the key to the solution of all problems by their "objective," "scientific" method.

Michael J. Moravcsik
University of Oregon

DISSENT

I strongly oppose the proposal to amend the APS constitution to permit adoption of resolutions on "any matter of concern to the Society." Without further restriction, the meaning of "a matter of concern" is simply the willingness of the majority of those voting to adopt the resolution. Thus there is no assurance that such resolutions would relate in any direct way to the furtherance of the aims of the society. On the contrary, the raising of non-scientific issues—even if not adopted—could have a deeply divisive effect on the society at a time when we need unity in the face of serious declines in relative numbers of students choosing physics for study and declining sources of support for basic research in physics.

Far short of resolutions on public matters not directly related to physics, there is serious doubt in my mind whether APS should even adopt positions on matters directly affecting physics. As examples one might suppose that APS might adopt a formal position with regard to the implementation of the recommendations of the Pake Report, might propose bilateral co-operation with the USSR in high-energy or space physics, might propose public support for programs to increase the numbers and improve the quality of high-school physics teachers. I would question whether even such topics as this should be considered in terms of membership resolutions in APS. The American Institute of Physics, with its magazine *Physics Today*, is an appropriate means for the expression of such views by physicists. Perhaps AIP should consider some means for ascertaining the statistical will of the physics community on specific issues directly related to the health of the science. But not APS.

The primary role of APS is to provide a medium of communication between physicists and to maintain the delicate balance between free communications—in print and orally—and the establishment of quality standards in that communication. Every physicist—not just the majority—must have complete confidence that the editors of our journals, the chairmen of our meeting sessions, the reviewers of our manuscripts and the councillors of our society are acting only on the basis of the highest standard of professional integrity. They must not be influenced by campaigns to try to produce a uniform opinion of physicists on the existence of quarks or time variation of the constant of gravitation.

Last Fall I heard Congressman Emilio Daddario tell the AIAA that they should take action as a society on behalf of publicly supported technical programs of great concern to them (such as space) as the Congress needs such publicly expressed views of subsections of the electorate on matters of direct self-interest. Arthur Kantrowitz and other officers of the AIAA correctly resisted this suggestion, for just the reasons I have given. Professional societies organized to run journals and meetings must not run the risk of alienating *any* professionally qualified member, for the societies have monopolies on the effective means of scientific communication.

I have circulated this letter to my colleagues in JILA and the Department of Physics and Astrophysics. Ten of them agree wholeheartedly and none of them has expressed disagreement with the views of this letter.

Lewis M. Branscomb
Joint Institute for Laboratory Astrophysics,
University of Colorado

SCHWARTZ'S REBUTTAL

I feel obliged to reply to some of the objections that have been raised against the proposed APS constitutional amendment in letters in this column and in the discussion at the Chicago meeting (reported in *Physics Today*, March, page 81). Although there have been a number of arguments of substance that the reader will evaluate for himself, it appears to me that several statements represent offscale reactions of alarm, and these should be put back into perspective.

Chief cause of the exaggerated responses was probably the unfortunate phrasing of the announcement presented by the editors of the *Bulletin*, "Should the APS broaden its purpose and aims to include discussion of public issues?" Many readers appear to have taken this question too literally and have drawn the false conclusion that some group is trying to turn the society into a wide-open political debating club. Problems of the pursuit of physics that are entangled with questions of public policy have been and will continue to be topics for discussion at APS meetings. An outstanding example was the stimulating session at the last annual meeting, "The Coupling of Physics and Society in the Seventies," sponsored by COMPAS, the AIP Committee on Physics and Society. The fear that such activities will grow to dominate the meetings is quite without basis; the APS council will certainly see to it that physics remains foremost.

A desire to revise the object and purpose of the society is another misrepresentation of the intent and substance of the amendment. My own expectation is that we can achieve a greater fulfillment of our stated purpose if we first disabuse ourselves of the notion that the activity of physics takes place in a vacuum without interacting strongly with the outside world. The society as a body is obliged to make judgments about whether or not particular matters of public policy impinge on "the advancement and diffusion of the knowledge of physics," and vice versa. What the proposed amendment does is to provide a mechanism whereby the membership at large can participate, when it cares to, in making these judgments. The habit has been to avoid controversial issues whatever their substance, and the one thing we do hope to change is this prudish habit.

Another element that has led to the extreme character of many letters and comments is the physicists' love of a good argument and the inclination to let a small prejudice lead one by logical progression to the most exaggerated conclusions. What we need most precisely in this area of the overlap of science with politics is a tempered weighing of facts, opinions and probabilities. The proposed amendment has been criticized for being less than perfectly constructed, and I will apologize for this failing even though I have yet to see any concrete suggestions that might improve it. If it is agreed that greater participation by the members is desired in studying how our professional concerns intermingle with public issues, then let us adopt the proposal as at least a reasonable instrument to start with. Then we can proceed with the experiment in earnest.

Charles Schwartz
University of California, Berkeley

WHAT WE ARE NOT AGAINST

In the wake of the campaign and vote on the Schwartz amendment, we hope that some muddy waters have become clearer to physicists (including ourselves). The proposal to have the American Physical Society pass resolutions on "any matter of concern," the letter debate in our columns and the oral debate in Chicago brought to the surface considerable turbulence that apparently has been bothering many for some time. We of *Physics Today* and the American Institute of Physics are concerned peripherally; only APS is directly involved. But we are learning while others learn, and we feel the whole controversy has thrown considerable light into the shadows. Perhaps the troubled waters look a bit smoother than before.

A few misconceptions remain, it appears, and some of them are about *Physics Today*. In the hope of greater clarity, we would like to try to remove some of them.

In the first place, we are frequently quoted as having taken a stand against the Schwartz amendment. The misconception arises because our rejection of the original Schwartz letter put the whole matter in motion. From that time on, to the extent that we were able, we treated proponents and opponents equally. Our December editorial expressed opinions against, but alongside them it expressed opinions for. We allowed proponents and opponents to inspect the draft of the editorial and modified it to include all the arguments they suggested. In handling the flood of letters initiated by our invitation and that of APS, we treated the sides in the same manner, generally allowing the proponents of change to tell their story first when we had to make a choice of order. If we had a bias and it showed, it showed despite our best efforts.

A second misconception is that *Physics Today* is against involvement and in favor of physics that is isolated from the problems that beset our world. We hope that anyone who thumbs through even our precampaign pages may find items to convince him otherwise: A Negro's view of the black man in physics, an Indian institute as an example of developed nations trying to help the developing, a story on pollution and crime, an editorial called "To Keep the Poor from Getting Poorer."

Let us say now that we are heartily in favor of physicists who want to find out what is wrong and set about correcting it. At the risk of playing devil's advocate, we express hearty sympathy with the undergraduate who turns aside from a physics major (and his father who leaves the physics faculty) because he feels a major in sociology or a job with the Peace Corps will help more poor and hungry people. And in all situations we feel that the physicist as individual and citizen (when *Physics Today* is not his medium) should be involved and concerned and contributing toward justice and happiness.

We try to further such causes, but we do so with two distinct reservations. One is that we feel *Physics Today* speaks mainly *to* the physicists not *for* them and only incidentally *from* them. In all our editorial judgments we keep firmly to our knowledge that our audience is professional physicists. If outsiders can gain something from what we write and print, so much the better. But we do not address ourselves to the outsiders. The second reservation is that we do not believe it is just or right or useful to take advantage of the "halo effect." If the physicist has gained unusual respect and influence by solving the problem of parity conservation, we are proud and happy. But we do not think that from this platform he should convince others and himself that he, more than the next man, has brains enough to see the facts and heart enough to understand their portent when facts and portent are not related to physics.

Modesty is a useful virtue when it preserves the effectiveness of the modest man.

The campaign has strengthened our conviction that the physicist is concerned and wants to be involved. Within our limitations, we hope to help him to be so.

R. Hobart Ellis, Jr.

2. Notice distributed calling for the formation of a new organization.

Announcing the Formation of a New Organization of Scientists Dedicated to Vigorous Social and Political Action

Over the past 25 years the scientific community has grown very large in numbers and in resources, but we have become complacent with our prestige and have failed to face the responsibilities created for us by our very material achievements. While we now see that many of the products of science and technology have become more a menace than a boon to the interests of human society, the dominant professional associations—such as the APS—have deliberately remained aloof from the desperate problems facing mankind today; and those few individual spokesmen and groups of scientists who do concern themselves with questions of government policy have failed to provide the bold leadership which is so badly needed. As scientists have become more and more dependent on the government for research funds and for their very livelihood, speaking out on public issues has been done more and more cautiously. We must therefore strive to regain our full intellectual and political freedom; and the very existence of this proposed organization will help strengthen each scientist's resolve.

An essential task is one of self-identification. We reject the old credo that "research means progress and progress is good." Reliance on such simplistic ethical codes has led to mistaken or even perverted uses of our scientific talents. (Consider the channeling of young scientists into weapons development work by the present complex of federal policies on education, funding and the draft.) As an antidote we shall establish a forum where all concerned scientists—and especially students and younger members of the profession—may explore the questions, Why are we scientists? For whose benefit do we work? What is the full measure of our moral and social responsibility?

As an ongoing organization we shall seek new and radical solutions for long range problems and immediate issues, and we shall press for effective political action. We shall work for change within our present affiliations (professional society, university, laboratory), but foremost we shall strive to present our opinions as an independent body of socially aware scientists free from the inhibitions which abound in the established institutions we now serve. We shall also seek to relate our activities to those of similar groups (radical caucuses) now forming in other professions.

3. Where to hold scientific meetings: statements by the Executive Board of the American Association of Physics Teachers and Professor Jay Orear (Cornell University); editorials and letters from *Nature*. (All reprinted with permission.)
Nature: "Governments as Patrons of Science,"
Vol. 224 (1969), 2;
Letter, Vol. 224 (1969), 93;
"When to Boycott," Vol. 226 (1970), 482;
Letter, Vol. 226 (1970), 573.

Two Statements

STATEMENT OF THE EXECUTIVE BOARD OF AAPT (Feb. 28, 1969)

Today no large city offers an atmosphere free from overt manifestations of deep social unrest. Activist groups and civic and state officials have spoken words and have taken actions that have alarmed and alienated thoughtful citizens. Who can predict which locale will be torn apart next week?

In the midst of such uncertainty, your elected officers still must schedule meetings years in advance. If they now must weigh the relative and variable socio-political demerits of possible meeting sites, whose criteria should they use?

At the present time AAPT is uniquely the national membership organization devoted to improvement of physics instruction and to furtherance of the appreciation of the role of physics in our culture. The latter clause has never heretofore been construed to sanction political or social action. If a new interpretation is called for, AAPT will probably change its character drastically. For one thing, the social and political views of candidates for officership will become important. If our professional concerns no longer dominate AAPT, what will hold it together?

The poll is going to tell us, we expect, that our membership is of two minds and that feelings on both sides can be strong indeed. The Executive Board will have to consider, as well, that our Chicago Section pleads officially for us to hold to our present meeting plan. It will have to cope with the question: By 1970, which will mean more, a gesture repudiating the events of 1968 or a meeting in a great center for physics? And the further question: Granted that physicists have no more right than anyone else to compartmentalize responsibilities and loyalties, is it possible for an association of physicists to be asked to serve too many purposes?

The Executive Board believes there is lasting value to AAPT and our decision, taking full account of the results of the poll, will be based on our best judgment of how to preserve and strengthen it.

STATEMENT OF JAY OREAR, CORNELL UNIVERSITY

The choice of our meeting site is, in a deep sense, much more than a political question. As a matter of principle the AAPT and the APS boycott hotels which discriminate according to race. For these same principles we should not meet in a place which many of our members feel has the policy of denying freedom of expression and using what these members consider as police-state tactics.

The AAPT is a professional society of science educators and by its very nature has obligations to the academic community at large. The AAUP, which traditionally sets the guide lines for the academic community has moved its annual meeting away from Chicago. Many other professional societies which had meetings scheduled for Chicago have followed this guide line. Some of these are the Societies of mathematics, psychology, philosophy, sociology, political science, modern language, and history.

The public image of American physics will be tarnished (especially in the eyes of those students we most wish to attract) if we and the APS stay in Chicago while so many other learned societies have already moved away.

Even if the APS should finally decide to remain in Chicago, we are still free to have our meeting in another location. Such an eventuality would most likely be of benefit to our Society. It might suddenly attract a large group of talented and idealistic physicists and graduate students to membership in our Society. It would demonstrate that we as teachers are able to reach our students—that we are responsive to the new wave of social concern sweeping through our young people.

Governments As Patrons of Science

The letter [following] from a group of distinguished molecular biologists raises several important and difficult questions about the relationships between private scientists and governments, their own or other people's. By all accounts, the matter was brought to a head by the summer school in Spetsai earlier this year, and in this sense the letter from Dr F. H. C. Crick and his associates should be read in conjunction with that from a group of students at the summer school at Spetsai earlier this year, and published a few weeks ago in *Nature* (**223**, 1186; 1969). The immediate problem is to define the circumstances in which scientists can with propriety agree to hold meetings in countries where governments have created an atmosphere which is hostile to intellectual life and possibly to science in particular, or where governments behave in a way which is simply illiberal by the standards of people outside. The general problem is to know how groups of scientists should regulate their relationships with governments and other official bodies in such a way as not to compromise themselves or be exploited for quite unscientific ends.

The signatories of the letter are right to start from the principle that science is international. It is no joke that young people in England are brought up with the tale of how Davy and Faraday went jaunting around the academies of Europe at the height of the Napoleonic Wars. It is also fair that the present Foreign Secretary of the Royal Society should be fond of telling public officials that the Royal Society thought it wise to have a Foreign Secretary before the British Government had appointed one. The point is simple but essential. The progress of science is necessarily fitful but certainly the more efficient and effective when scientists everywhere are able to contribute towards it. The fact that people live in different countries is often not so much an impediment as an incentive to collaboration—people often learn more from each other when they work in different environments. This, no doubt, is a part of the reason why the Cern laboratory at Geneva has been such a success. This, too, is why research programmes such as that which has taken a British group to the Russian thermonuclear laboratories turn out to be especially stimulating. These, however, are only conspicuous ways in which science is known to be larger than the sum of its national parts. The remarkable way in which the scientific communities of North America and Western Europe have been cemented together, since the war, simply by trans-Atlantic visiting is a proof that science could not continue as it has become without the free movement of scientists. If the international character of science may occasionally help to smooth the way for other kinds of international relations, that is a valuable but uncovenanted benefit.

In the circumstances, it is also proper of the signatories of the letter to imply that there was nothing reprehensible in the holding of the summer school at Spetsai. Nobody seems to have much love for the undeserving Government of Greece, but it also seems to be the case that the Spetsai conference owed almost nothing to the colonels. It is important that money is often a simple measure of propriety in circumstances like these. If a government has paid the bill for the holding of a conference, it can rightly take some of the credit for what success there may be to boast about. So does it follow that all public money is tainted? And should people travel to conferences on air tickets paid for out of their own pockets? Conferences would quickly die out if that rule were applied. The test is that public money should not be spent directly by governments on the patronage of science but, rather, given first of all to learned societies or other organizations which can be counted on to spend what funds there are without prejudice. And here, of course, it is essential that those whose presence at a conference is considered desirable should be free to attend. Briefly, a government wishing to have a conference on its doorstep must reckon to provide visas for all those whom the organizers would invite. Since the war, this has been a recurring source of trouble, and there is even now no means of knowing how successful would be some kinds of international conferences in Brazil. And would Cuban scientists be welcome to come and go from international conferences in the United States? This is a matter to which ICSU should give its attention. But in circumstances in which the freedom to come and go is established, there may be great advantages in holding conferences in hostile environments, not the last of which is that of lifting some of the sense of isolation from which members of an oppressed intellectual community must suffer.

But what if a government should be free with visas and even tolerant of the ways in which the recipients of conference money wish to spend their funds? Is it necessary for scientists still to worry about the reflected glory which a meeting such as that on Spetsai is likely to bring? The trouble here, of course, is that the bodies of scientists which are usually concerned with the organization of conferences are not especially able at making collective decisions about the worthiness of governments. Committees are never quick and usually divided on matters not directly relevant to their existence. In the circumstances, it is unrealistic to expect that the organizers of conferences can discriminate between governments, picking the goats from the sheep. The best they can do, and the most that can be expected of them, is so to arrange their affairs that no direct benefits accrue to any government. This, to be sure, is a fierce doctrine. It would, for example, suggest that learned societies should not accept public money earmarked for purposes which are specified in detail. A general fund for conferences would be permissible, but not a gift of money for a specific meeting. Present practice is more lax than this ascetic doctrine, and the signatories of the letter [following] would not go as far, but there is a case for thinking that a substantial measure of self-denial by the scientific community is a part of the price which must be paid to allow the scientific community fairly to cherish its international character.

But if scientists hold governments at arm's length, is there not a danger that the coolness will be reciprocated? This will no doubt be the fear in the minds of many treasurers of learned societies, not to mention many would-be participants at international conferences. It is most probably an unrealistic fear. By now, governments have come to appreciate the value of a genuinely independent scientific community. Raising money without strings attached is a much easier matter than before the Second World War. Indeed, many governments have come to appreciate the ways in which independent learned societies can accomplish things which are beyond the reach of governmental organizations. The European exchange scheme which is being administered by the Royal Society (where the United Kingdom contribution is concerned) and by other European academies is a ready illustration of this principle.

The moral of the disagreement about Spetsai is that societies and the scientific community which they represent should be less afraid than in the past of seeking and proclaiming financial independence.

International Conferences

Sir,—There have recently been several occasions when scientists have refused to attend certain meetings because they have objected to some of the actions of the government of the country in which the meeting was being held. It seems likely that such situations will recur in the future. These and similar problems arise in international scientific relations because of the profound division of the world into ideologically, politically and socially different systems. We feel it important that the issues involved should be discussed by the scientific community. We do not aim to lay down in detail the judgment to be made in any particular instance, but to see if some measure of agreement can be obtained on a few basic principles and rules which could serve as guidelines for making decisions. In the last analysis we believe that such decisions must remain the personal responsibility of the individual scientist.

The great majority of scientists would agree that: (1) Science is international. Free and constant intellectual communication between scientists is essential for the health of science, and frequent direct personal contact is very desirable. (2) The scientific community should not be divided because of nonscientific issues, if such division can be avoided.

Nevertheless, certain national aspects of science are unavoidable. The funds for most scientific research and teaching are provided from national sources. Meetings must take place in some particular country or other, unless we are only to meet on the high seas. In addition, at the present time we have to face the following facts. (*a*) Some countries put more or less severe restrictions on foreign travel by their own citizens, and/or on visits by citizens of certain other countries. (*b*) Scientists and scholars have (quite recently in some countries) been dismissed or imprisoned without fair and public enquiry or trial. (*c*) In some countries the public discussion of certain scientific ideas may be difficult, if not actually restricted, because they are considered to contradict or question the official philosophy which forms the basis of the social system. (*d*) Because many meetings, congresses, or even individual visits are officially sponsored by governments,[1] some scientists feel that their participation in such activities could be construed as implying their approval of a system or policy which in fact they strongly dislike. They feel morally bound to decline certain invitations even though their acceptance might contribute to fruitful scientific communication.

In making a decision on these matters, a scientist, we feel, should not be primarily concerned with the question as to which course of action might best serve his own public image. Nor should he let his personal sympathy or antipathy towards a regime or its ideology *automatically* decide whether he should accept a foreign invitation or not. As a citizen he may hold, and express, his opinions of a certain social system, of its ideology or of the policy of its government. As a scientist, and when confronted with a specific issue whether or not to accept an invitation directly related to his field of activity, his major concern should be to try to make the decision which, in his opinion, would best serve the international scientific community and encourage freedom of expression and communication. He should therefore not allow his authority or prestige to serve, implicitly or otherwise, the propaganda of any regime or organization responsible for putting restrictions *of any kind* on the freedom of communication between scientists or on academic freedom in general. At the same time, he will not want to penalize scientists of another country because of the oppressive or restrictive policy of their government, of which, in many cases, they are the first victims and the strongest opponents.

It would follow, therefore, that invitations sponsored or honours bestowed by a government responsible for any sort of restriction on the freedom of science and scientists should be declined. The reason for the refusal should be clearly stated, and refer not so much to the general policy of the government as to its attitude towards its own scientific and academic community. Private invitations[2] by individual colleagues, universities or institutes, by contrast, could be accepted, and should whenever possible be turned into an occasion for reaffirming publicly the unity of the scientific community and its opposition to any ideological or political oppression.

Attendance at a particular meeting in another country might, in suitable cases, be made dependent on a set of conditions aimed at making the meeting an open and unofficial one. Such conditions might be: (1) the government would not prevent the attendance of any *bona fide* scientist at the meeting; (2) the government would not make any political propaganda about the meeting being held in their country; (3) officials of the government would not address the meeting; (4) direct[3] financial support of the meeting by the government would not be acceptable,[4] except perhaps for optional cultural activities which individual scientists could feel free to refuse if they so wished.

We fully realize that, while these principles and rules seem clear and simple enough, it may be difficult in many particular instances to decide just where to draw a line and how to make one's own attitude known without allowing it to be unduly exploited and without endangering certain colleagues. In spite of these difficulties we feel that a wide consensus, within the scientific community, in favour of actively defending these principles on every possible occasion is likely in the long run to serve not only the development of science but also the wider cause of civil liberties and human rights.

In any case, we hope that our suggestions will provoke a wider discussion of the issues involved and of the correct course of behaviour to be followed. We suggest that any scientist who agrees or otherwise with our general position might usefully send a postcard or letter to the Editor of *Nature* to that effect.

Yours faithfully,
F. H. C. Crick
J. C. Kendrew
M. F. Perutz
F. Sanger
Medical Research Council
Laboratory of Molecular Biology,
Cambridge
Jacques Monod
Francois Jacob
Institut Pasteur,
Paris, France
André Lwoff
Institut de Recherches Scientifiques
sur le Cancer,
B.P. No. 8,
94, Villejuif, France

1. We would also include political or military organizations supported by several governments. The discussion might reasonably be extended to organizations within a country, such as military establishments, commercial firms, and so on, but to avoid complicating the issue we suggest that these cases be left aside for the moment.

2. We realize that there are some countries where all such private initiative is controlled and any invitation would have to be considered an official one.

3. It may be difficult in some cases to decide whether the support is "direct." In assessing this it would seem sensible to consider whether there are any strings attached to the granting of the money, or whether the money is allocated on a strictly scientific basis, without any political or military considerations.

4. Scientists at the present time appear to be divided on the ethical issue of whether one should accept money from a government of which one disapproves. Some feel strongly that money should not be accepted. Others argue that such financial contributions, though small, will, if anything, weaken the organization which makes them. Because we believe that even after debate there will always be a substantial fraction of scientists who are against accepting such money, we suggest that no useful purpose will be served by publicly debating this particular ethical point in this context.

When to Boycott

Mr Nigel Calder, a distinguished writer on scientific subjects, seems sadly to lack a sufficient sense of the ironical. This, at least, is the most charitable interpretation of his account, in the *New Statesmen* on March 20, of the comments which have appeared in *Nature* in the past few weeks on the dismissal of Professor Otto Wichterle from his post as director of the Institute of Macromolecular Chemistry in Czechoslovakia. To begin with, Mr Calder has dignified as "calling for a boycott of two conferences in Czechoslovakia" the wry comment in "Miscellaneous Intelligence" (*Nature*, **225**, 120; 1970) that "those who had intended to accept Professor Wichterle's invitations may think twice about doing so should it prove that their host is unable to reciprocate their visit". Then he seems to have read Professor Wichterle's moving comment (*Nature*, **225**, 773; 1970) as a "bid to protect the conferences from a Western boycott." This is not the place to mull over again the circumstances which led to the dismissal of Professor Wichterle, but the issue of when and how professional scientists should respond to oppressive acts against professional organizations by governments elsewhere is important and, unhappily, increasingly topical.

The most obvious but the most important thing to say is that there can be no simple rules to guide the way in which scientists respond to the most frequent challenges to the conscience—invitations to attend meetings or to give lectures. Boycotts have the obvious disadvantage that the people who suffer first are the beleaguered scientists, not their governments—this is one of the points which Professor Wichterle made. But it is also inconceivable that professional scientists could ever be sufficiently united in their reactions to particular governments for them to exert a decisive influence on political events. Indeed, the first objective should be somehow to arrange that science should in present circumstances as in the past be able to transcend political differences between nations and even individuals—a doctrine which does not imply that political differences are unimportant but, rather, that common interests in science can contribute significantly to the resolution of differences. One important corollary is that there should be a carefully cultivated detachment of learned societies from the governments of the nations which they inhabit. Only then can potential participants in conferences be sure that their presence will not unreasonably add to the prestige of governments with which they may be out of sympathy. Another is that learned societies, and all organizers of conferences in particular, should be fierce in their insistence that governments should not interfere with the freedom of scientists to come and go on legitimate business.

The virtues of these objectives are plain enough; their feasibility is another matter, but that is no reason for indifference or even despondency. The issue of free travel to scientific conferences has been alive for several years, with governments of all kinds playing the fool from time to time. In the West, there is continuing and needless trouble about visas from people in East Germany. Visits to Eastern Europe and the Soviet Union are still conducted at a more or less governmental level, with academies of science involved as travel agents if nothing more. Although mainland China is still almost entirely cut off from everywhere, there has been a good deal of improvement since the fifties, when there were times when fully international conferences seemed impossible. Even so, there is a long way to go. The fact that many people living in Eastern Europe and the Soviet Union, and India for that matter (see *Nature*, **225**, 116; 1970), are not free to attend scientific conferences to which they are invited without the consent of their governments is not a reason for any kind of boycott, but unreasonable denial of permission is an affront against the free traffic in scientific ideas on which everybody depends. To say this is not to imply that scientists should be immune from the restrictions which prevent other kinds of employees from swanning around the world whenever there is an occasion to do so, which means that many denials of permission may be sensible enough and which certainly implies that each case must be judged on its merits. But there is a strong case for a systematic study of these problems, possibly as an extension of the good work which the ICSU committee under Professor N. Herlofson has already undertaken. With persistence, it should be possible to shame most governments into sensible behaviour.

The broader issue of the independence of the learned societies from their governments raises more serious difficulties. The objective is that learned societies should be so detached from the policies of their governments that their work is not compromised by the policies which their governments pursue. One crude but effective test is whether a society operating in a country with a government of which some people abroad disapprove can organize a meeting on its home ground without seeming to be an instrument of distasteful public policies. The dependence or otherwise of the society on public funds is less important than its actual freedom to spend its money as it chooses. It is fair but sad to say that even the learned societies, which are fierce (and just) defenders of intellectual independence, are too ready on occasions to represent their governments or, worse still, to be patronized by them. Some of the most respectable societies are so dependent on government money for essential functions that they must think twice before biting the hand that feeds them. And some societies are hardly respectable at all.

In the long run the best guarantee of independence is a greater degree of financial autonomy, which is one important matter to which the scientific community as a whole should pay attention. But where international societies, and ICSU itself, are concerned, the hazards lie just beneath the surface. It will be a long job to get rid of them, and so embody the independence of the scientific community in thoroughly international organizations. In practical terms, however, this is the direction in which Professor Wichterle's case points.

When to Attend

Sir,—In a recent issue of *Nature* (**223**, 1186; 1969) a large number of non-Greek scientists who attended the fourth NATO Advanced Study of Molecular Biology held in Spetsai stated that they deplored the Greek military government but felt that it was necessary to attend this meeting so that the Greek intellectual community does not become isolated and cut off from free interchange of ideas. More recently (*Nature*, **224**, 93; 1969) some distinguished English and French scientists (including Dr F. H. C. Crick, who was the chairman of the Spetsai meetings) gave a long analysis of the guide lines that members of the scientific community may use to help them decide which international conferences they should attend. An editorial in the same issue (*Nature*, **224**, 2; 1969) further elaborates on the international nature of the scientific community and the independence of scientists to meet wherever they wish without having to decide on the "worthiness" of governments—provided that the scientific societies are financially independent. If a military junta (such as that in Greece) has censored the freedom of expression and has cancelled the most basic constitutional rights of its own citizens, why assume that conditions of attendance would be adhered to? Both letters profess a deep interest in the scientists who live in a country run by a dictatorship and a strong need to communicate with them. However, both letters neglect to point out a very crucial aspect of such meetings, and that is whether the attendance of scientists from the host country is dependent on their government's clearance. While well-known scientists might be missed, promising young investigators who are not well-known outside their own country could be screened and their absence at the meeting would go unnoticed.

No set of principles or guide lines can be used in deciding whether to attend a scientific meeting held in a country under an unprincipled dictatorship. Perhaps only one question is important, and that is whether, by boycotting or, better still, cancelling such a meeting, this action will contribute to the unpopularity of a regime and its eventual change. In small countries, such as Greece, this is certainly very important since critical foreign opinion, including that of the international scientific community as well as economic boycott (decrease of tourism, etc.), will play an important part in the eventual removal of the military junta.

Yours faithfully,
George D. Pappas
Albert Einstein College of Medicine,
Yeshiva University,
Bronx, NY 10461

bibliography

1. Books

Listed below, in several broad categories, are books that may make interesting reading or provide a start for more detailed studies. Some have already been mentioned in the text. Wherever possible, paperback editions have been listed and so indicated.

(A) TOPICS IN THE AREA OF SCIENCE AND CONTEMPORARY SOCIETY; SOCIAL AND POLITICAL ASPECTS

Applied Science and Technological Progress, Report to the Committee on Science and Astronautics, by the National Academy of Sciences, to the U.S. House of Representatives. Washington, D.C.: U.S. Government Printing Office, 1967 (p/b).

Bernal, J. D., *The Social Function of Science*. Cambridge, Mass.: M.I.T. Press, 1967. (p/b reissue of a classic treatise, first published in 1939.)

Brooks, Harvey, *The Government of Science*. Cambridge, Mass.: M.I.T. Press, 1968.

Brown, Martin, ed., *The Social Responsibility of the Scientist*. New York: The Free Press, 1971 (p/b). (Mainly concerned with documenting problems that arise from the practice and application of science. Useful list of references.)

Cleaveland, Frederic N., *Science and State Government*. Chapel Hill, N. C.: The University of North Carolina Press, 1959.

Commoner, Barry, *Science and Survival*. New York: Viking Press, 1966. (An original and critical appraisal of the role of the scientist. Useful notes and references; p/b.)

Crowther, J. G., *Science in Modern Society*. London: The Cresset Press, 1967. (Collected essays, mainly on the British scene.)

Dupre, J. Stefan and Sanford A. Lakoff, *Science and Nation*. Englewood Cliffs, New Jersey: Prentice-Hall, Inc., 1962 (p/b).

Goldsmith, M. and A. Mackay, *The Science of Science*. London: Penguin Books, 1964 (p/b). (Commentaries on the twenty-fifth anniversary of publication of Bernal's book—see above.)

Greenberg, Daniel S., *The Politics of Pure Science*. New York: New American Library, 1967. (Comments on the scientific politics; Greenberg's column in *Science* did much to enliven and alert the scientists' awareness for several years.)

Grodzins, Morton and Eugene Rabinowitch, eds., *The Atomic Age*. New York: Simon and Schuster, 1965. (Collected essays from the *Bulletin of the Atomic Scientists*, 1945–62; p/b.)

Fischer, Robert B., *Science, Man and Society*. Philadelphia: W. B. Saunders Co., 1971. (Concerned with many of the topics covered in the present book. A good list of references; p/b.)

Haberer, Joseph, *Politics and the Community of Science*. New York: Van Nostrand Reinhold, 1969. (Interesting commentary and analysis by a political scientist; good references; p/b.)

Hagstrom, Warren O., *The Scientific Community*. New York: Basic Books, Inc., 1965. (Enquiry into the social organization of science based on interviews by a sociologist.)

Joravsky, David, *The Lysenko Affair*. Cambridge, Mass.: Harvard University Press, 1970. [See also *Scientific American*, (November 1962), p. 41.)]

Karplus, Robert, *Physics and Man*. New York: W. A. Benjamin, Inc., 1970. (Readings, some on methods of science, others on relation to society; p/b.)

Logsdon, John M., *The Decision to go to the Moon: Project Apollo and the National Interest*. Cambridge, Mass.: M.I.T. Press, 1970.

March, Michael S., *Federal Budget Priorities for Research and Development*. Chicago: University of Chicago Press, 1970. (Concise tabulation of distribution and trends in R and D Funds; p/b.)

Medvedev, Zhores A., *The Rise and Fall of T. D. Lysenko*, transl. by I. Michael Lerner, with editorial assistance of Lucy G. Lawrence. New York: Doubleday Anchor Books, Doubleday and Co., 1971. (Remarkable documentation of an amazing era, by a prominent Russian biochemist; p/b.)

Mesthene, E. G., *Technological Change*. New York: Mentor Books, New American Library Inc., 1970 (p/b.)

Morison, Elting S., *Men, Machines and Modern Times*. Cambridge, Mass.: M.I.T. Press, 1966. (Essays on the introduction of new technologies in the United States in the nineteenth century; p/b.)

Nelkin, Dorothy, *Nuclear Power and Its Critics*. Ithaca, New York: Cornell University Press, 1971. (Description of recent case in which scientists and citizens combined to prevent siting of a nuclear power plant on Cayuga Lake; p/b).

Obler, Paul C. and Herman A. Estrin, eds., *The New Scientist*. New York: Doubleday Anchor Books, Doubleday and Co. Inc., 1962. (Collected essays; p/b.)

Orlans, Harold, ed., *Science Policy and the University*. Washington, D.C.: The Brookings Institution, 1968 (p/b).

Piel, Gerard, *Science in the Cause of Man.* New York: Vintage Books, 1964. (Essays by the publisher of the *Scientific American*; p/b.)

Price, Don K., *Government and Science.* London: Galaxy Books, Oxford University Press, 1962 (p/b.)

Price, Don K., *The Scientific Estate.* Cambridge, Mass.: Harvard University Press, 1965 (p/b.)

Price, D. J. de Solla, *Little Science, Big Science.* New York: Columbia University Press, 1963. (Famous study of some facets of the operation and growth of science; p/b.)

Rose, Hilary and Stephen Rose, *Science and Society.* London: Allen Lane—The Penguin Press, 1969. (By two of the founding members of the British Society for Social Responsibility in Science.)

Sakharov, Andrei D., *Progress, Coexistence and Intellectual Freedom.* London: Andre Deutsch, 1968. (Essays by a leading Russian physicist.)

Shils, Edward, ed., *Criteria for Scientific Development: Public Policy and National Goals.* Cambridge, Mass.: M.I.T. Press, 1968. (Selected articles from *Minerva:* see listed among periodicals, below; p/b.)

Skolnikoff, Eugene B., *Science, Technology and American Foreign Policy.* Cambridge, Mass.: M.I.T. Press, 1967 (p/b.)

Smith, Alice Kimball, *A Peril and a Hope.* Chicago: University of Chicago Press, 1970. (Study of the atomic scientists' movement, 1945–47; p/b.)

Snow, C. P., *Science and Government.* New York: Mentor Books, 1962. (Famous essay on science and politics in Britain; p/b.)

Snow, C. P., *The Two Cultures and a Second Look.* New York: Mentor Books, The New American Library, 1963. (The publication of the original essay in 1959 occasioned some intemperate replies but also much good discussion of the issues raised by Snow; p/b.)

Strickland, Donald A., *Scientists in Politics: The Atomic Scientists Movement 1945–6.* Lafayette, Indiana: Purdue University Studies, 1968.

Washington Colloquium on Science and Society. Vol. 1, 1964–5, ed. by S. Frederick Seymour; Vol. 2, 1965–6, ed. by Morton Leeds. Baltimore, Maryland: Mono Books Corporation, 1967.

Weinberg, Alvin M., *Reflections on Big Science.* Cambridge, Mass.: M.I.T. Press, 1967. (Thoughtful essays on some aspects of modern science; p/b.)

(B) HISTORICAL STUDIES ON THE ROLE OF SCIENCE AND TECHNOLOGY IN SOCIETY

Armytage, W. H. G., *A Social History of Engineering.* Cambridge, Mass.: M.I.T. Press, 1961.

Ashton, T. S., *The Industrial Revolution.* London: Galaxy Books, Oxford University Press, 1964 (p/b).

Basalla, George, William Coleman, and Robert H. Kargon, *Victorian Science.* New York: Anchor Books, Doubleday and Co., Inc., 1970. (Nineteenth-century Presidential addresses to the British Association for the Advancement of Science; p/b.)

Dampier, W. C., *A History of Science.* Cambridge: Cambridge University Press, 1966. (Older classic, good as an introduction.)

Marsak, L. M., ed., *The Rise of Science in Relation to Society.* London: The Macmillan Co., 1964 (p/b).

Merton, Robert K., *Science, Technology and Society in Seventeenth Century England.* New York: Harper Torchbooks, Harper and Row, 1970. (Reprint of 1938 essay, still of considerable interest; p/b.)

Musson, A. E. and Eric Robinson, *Science and Technology in the Industrial Revolution.* Toronto: University of Toronto Press, 1969.

Needham, Joseph, *The Grand Titration: Science and Society in East and West.* Toronto: University of Toronto Press, 1969. (Reprint of many essays, by the distinguished historian of science in the Far East.)

Schofield, Robert E., *The Lunar Society of Birmingham.* London: Oxford University Press, 1963. (Account of the meetings and work of an influential group of men, at the start of the Industrial Revolution.)

Singer, Charles, E. J. Holmyard, A. R. Hall, and Trevor I. Williams, eds., *A History of Technology* (five volumes). London: Oxford University Press, 1954–59.

White, Lynn, Jr., *Medieval Technology and Social Change.* London: Oxford University Press, 1962 (p/b).

(C) MOSTLY ON PHILOSOPHIC AND HUMANISTIC ASPECTS OF SCIENCE

Arons, A. B. and A. M. Bork, eds., *Science and Ideas.* Englewood Cliffs, New Jersey: Prentice-Hall, Inc., 1964 (p/b).

Bronowski, J., *Science and Human Values.* New York: Harper Torchbooks, Harper and Row, 1959 (p/b).

Lindsay, R. B., *The Role of Science in Civilization.* New York: Harper and Row, 1963.

Vavoulis, A. and A. W. Culver, eds., *Science and Society.* San Francisco: Holden-Day Inc., 1966 (p/b).

(D) GALILEO AND THE CHURCH

Brecht, Bertolt, *Galileo.* (English version by Charles Laughton, introduction by Eric Bentley.) New York: Grove Press Inc., 1966 (original German version, 1940) (p/b).

Drake, Stillman, *Galileo Studies.* Ann Arbor, Michigan: University of Michigan Press, 1970.

Geymonat, Ludovico, *Galileo Galilei.* New York: McGraw-Hill Book Co., 1965 (p/b).

Koestler, Arthur, *The Sleepwalkers.* New York: Grosset and Dunlap, 1963 (p/b).

McMullin, Ernan, ed., *Galileo: Man of Science.* New York: Basic Books Inc., 1967.

de Santillana, Giorgio, *The Crime of Galileo.* Chicago: University of Chicago Press, 1955 (p/b).

(E) THE OPPENHEIMER CASE

Chevalier, Haakon, *Oppenheimer: The Story of a Friendship.* New York: Pocket Books, Inc., 1966 (p/b).

Davis, Nuel Pharr, *Lawrence and Oppenheimer*. Greenwich, Conn.: Fawcett Publications, Inc., 1968 (p/b).

In the Matter of J. Robert Oppenheimer. Transcript of a Hearing before Personnel Security Board. Washington, D.C.: U.S. Government Printing Office, 1954.

Kipphardt, Heiner, *In the Matter of J. Robert Oppenheimer*. New York: Hill and Wang, 1967. (A play, using actual extracts from the Hearings for the dialogue; p/b.)

Stern, Philip M., with collaboration of Harold P. Green, *The Oppenheimer Case*. New York: Harper and Row, 1969.

(See also sections in Grodzins and Rabinowitsch, *The Atomic Age*, and Haberer, *Politics and the Community of Science*, listed under (*a*) above.)

(F) METHODS AND OPERATION OF SCIENCE

Baker, Jeffrey and Garland E. Allen, *Hypothesis, Prediction and Implication in Biology*. Reading, Mass.: Addison-Wesley Publishing Co., 1968. (Contains a good set of readings on a scientific controversy involving racial differences and suggested reasons; p/b.)

Brody, Baruch A., ed., *Readings in the Philosophy of Science*. Englewood Cliffs, New Jersey: Prentice-Hall, Inc., 1970.

Brody, B. and N. Capaldi, eds., *Science: Men, Methods, Goals* (A Reader: Methods of Physical Science). New York: W. A. Benjamin, Inc., 1968 (p/b).

Kuhn, Thomas S., *The Structure of Scientific Revolutions*. Chicago: University of Chicago Press, 1962. Second edition, with postscript, 1970. (A classic study; p/b.)

Medawar, P. B., *The Art of the Soluble*. London: Methuen and Co., 1967. (Trenchant views expressed in brief essays; p/b.)

Taton, R., *Reason and Chance in Scientific Discovery*. New York: Science Editions, Inc., 1962 (p/b).

Toulmin, Stephen, *The Philosophy of Science*. New York: Harper Torchbooks, Harper and Row, 1960. (Good introduction to the subject; p/b.)

Walker, Marshall, *The Nature of Scientific Thought*. Englewood Cliffs, New Jersey: Prentice-Hall, Inc., 1963 (p/b).

Ziman, John, *Public Knowledge*. Cambridge, England: Cambridge University Press, 1968. (Excellent study and commentary on some of the ways of science; p/b.)

(G) INTRODUCTORY BOOKS ON THE QUANTITATIVE ASPECTS OF SCIENCE

Baird, D. C., *Experimentation*. Englewood Cliffs, New Jersey: Prentice-Hall, Inc., 1962 (p/b).

Dantzig, Tobias, *Numbers: The Language of Science*, 4th ed., New York: Doubleday Anchor Books, Doubleday and Co., 1954 (p/b).

Moroney, M. J., *Facts from Figures*. London: Penguin Books Ltd., 1951. (An outstanding introduction to the ideas of probability and statistics, with a light touch; p/b.)

Polya, G., *How To Solve It*. New York: Doubleday Anchor Books, Doubleday and Co., 1957 (p/b).

2.   Resource Letters in Physics

These have appeared in the *American Journal of Physics*, as part of a continuing series, in which extensive bibliographies are compiled, together with brief comments and rating of references as elementary, intermediate, and advanced. The Resource Letters listed below have a far wider appeal than simply the physics community.

Bork, A. M. and A. B. Arons; "Collateral Reading for Physics Courses," **35**, No. 2 (1967).

Davenport, William H., "Technology, Literature and Art Since World War II," **38**, No. 4 (1970).

Nicolson, Marjorie, "Science and Literature", **33**, No. 3 (1965).

3.   Periodicals

All of these contain much of interest to the nonexpert; depending on their style, they may contain news and commentary, review articles on topical subjects, and book reviews. Some will also feature more technical articles. Listed with publisher and frequency of appearance.

Advancement of Science. (British Association for the Advancement of Science; 6 issues per year).

American Journal of Science (Kline Geology Laboratory, Yale University; monthly).

American Scientist (The Society of the Sigma Xi; bimonthly).

Bulletin of the Atomic Scientists: Science and Public Affairs (Educational Foundation for Nuclear Science, Chicago; monthly).

Chemical and Engineering News (American Chemical Society; weekly).

Chemistry and Industry (Society of Chemical Industry; weekly).

Chemistry in Britain (Royal Institute of Chemistry; monthly).

Daedelus (American Academy of Arts and Sciences; quarterly).

Endeavour (Imperial Chemical Industries, England; monthly).

Environment (Scientists' Institute for Public Information; previously published by St. Louis Committee for Environmental Information; ten issues per year).

Geotimes (American Geological Institute; monthly).

Impact of Science on Society (U.N.E.S.C.O.; quarterly).

Isis (History of Science Society; quarterly).

Minerva (Macmillan Journals Ltd., London; quarterly).

Natural History (American Museum of Natural History; monthly).

Nature (Macmillan Journals Ltd., London; 3 issues per week: one physical science, one biological science, one mainly news and reviews).

New Scientist (New Science Publications, London; weekly).

Philosophy of Science (Philosophy of Science Association; quarterly).

Physics Today (American Institute of Physics: monthly).

Quarterly Review of Biology (American Society of Naturalists; quarterly).

Science (American Association for the Advancement of Science; weekly).

Science News (Science Service Inc.; weekly).

Scientific American (Scientific American Inc.; monthly).
Science Policy Reviews (Batelle Memorial Institute; quarterly).
Sky and Telescope (Sky Publishing Co.; monthly).
Technology and Culture (Society for the History of Technology; quarterly).
Technology Review (M.I.T. Press; 9 issues per year).

4. Recent Publications

The following books have appeared too recently to have received mention in the body of the text:

Ben-David, Joseph: *The Scientist's Role in Society*; Prentice-Hall Inc., Englewood Cliffs, New Jersey; 1971. (p/b). Scholarly survey, tracing historical development of the scientist's role and its evolution.

Fuller, Watson (ed.): *The Social Impact of Modern Biology*; Routledge and Kegan Paul, London; 1971. (p/b). Papers and discussions presented at an international conference held in London in November, 1970, organized by the British Society for Social Responsibility in Science.

Ravetz, J. R.: *Scientific Knowledge and its Social Problems*; (Oxford University Press, Oxford; 1971. A major contribution to the contemporary study of this subject.

index

index